The Transactional Model *of* Development

The Transactional Model *of* Development

How Children and Contexts Shape Each Other

Edited by **Arnold Sameroff**

American Psychological Association • Washington, DC

Published by
American Psychological Association
750 First Street, NE
Washington, DC 20002
www.apa.org

To order
APA Order Department
P.O. Box 92984
Washington, DC 20090-2984
Tel: (800) 374-2721; Direct: (202) 336-5510
Fax: (202) 336-5502; TDD/TTY: (202) 336-6123
Online: www.apa.org/books/
E-mail: order@apa.org

In the U.K., Europe, Africa, and the Middle East, copies may be ordered from
American Psychological Association
3 Henrietta Street
Covent Garden, London
WC2E 8LU England

Typeset in Goudy by Circle Graphics, Columbia, MD

Printer: McNaughton & Gunn, Saline, MI
Cover Designer: Mercury Publishing Services, Rockville, MD
Technical/Production Editor: Harriet Kaplan

Library of Congress Cataloging-in-Publication Data

The transactional model of development : how children and contexts shape each other / edited, Arnold Sameroff. — 1st ed.

p. cm.

Includes bibliographical references and index.

ISBN-13: 978-1-4338-0467-0

ISBN-10: 1-4338-0467-0

1. Genetic psychology. 2. Developmental psychology. 3. Nativism (Psychology) 4. Nature and nurture. I. Sameroff, Arnold J.

BF711.T73 2009
155.4—dc22

2008049044

British Library Cataloguing-in-Publication Data

A CIP record is available from the British Library.

Printed in the United States of America
First Edition

CONTENTS

CONTRIBUTORS

J. Lawrence Aber, New York University, New York, NY
Marc H. Bornstein, National Institute of Child Health and Human Development, Bethesda, MD
Daphne Bugental, University of California, Santa Barbara
Margaret Clements, New York University, New York, NY
Carol McDonald Connor, Florida State University, Tallahassee
Kenneth A. Dodge, Duke University, Durham, NC
Alan Fogel, University of Utah, Salt Lake City
Reid Griffith Fontaine, University of Arizona, Tucson
Mary Gauvain, University of California, Riverside
Elizabeth T. Gershoff, University of Michigan, Ann Arbor
Richard Gonzalez, University of Michigan, Ann Arbor
Heather E. Gross, University of Pittsburgh, Pittsburgh, PA
Leon Kuczynski, University of Guelph, Guelph, Ontario, Canada
Erika S. Lunkenheimer, Colorado State University, Fort Collins
Michael J. MacKenzie, Columbia University, New York, NY
Susan C. McDonough, University of Michigan, Ann Arbor
Kristin L. Moilanen, West Virginia University, Morgantown
Frederick J. Morrison, University of Michigan, Ann Arbor

Sheryl L. Olson, University of Michigan, Ann Arbor
C. Melanie Parkin, University of Guelph, Guelph, Ontario, Canada
Arnold Sameroff, University of Michigan, Ann Arbor
Daniel S. Shaw, University of Pittsburgh, Pittsburgh, PA

PREFACE

How does one explain developmental surprises when children who are deaf and blind grow to become respected public figures, such as Helen Keller, or when the children of distinguished, affluent families become substance abusers and school dropouts? Not by appeals to theories that explain everything in terms of nature and nurture. What was and is still necessary is to understand how people and their environments work together to reach developmental success or failure.

"He changes the baby and the baby changed him" was a headline in the April 26, 2004, issue of *People* magazine. The idea that children affect their families and their families affect them is now common sense for developmental scientists and a large segment of the public. Despite the popularity of this idea, such reciprocal processes are rarely demonstrated in psychological research. The transactional model[1] that incorporates such bidirectional influences has been around for almost 40 years; however, researchers are still daunted by designing and analyzing transactional studies. Only now is behavioral science

[1]Sameroff, A. J., & Chandler, M. J. (1975). Reproductive risk and the continuum of caretaking casualty. In F. D. Horowitz, M. Hetherington, S. Scarr-Salapatek, & G. Siegel (Eds.), *Review of child development research* (Vol. 4, pp. 187–244). Chicago: University of Chicago.

catching up with the theory and overcoming the resistance to incorporating dynamic methodologies into a discipline that had been devoted to viewing individuals in isolation as a constellation of unchanging traits.

Forty years ago, the statistical method of choice was the analysis of variance with an emphasis on single variables that produced group differences on single dimensions. The transition to regression models was a major step toward the acceptance of some complexity in outcomes in which the effects of several variables could be examined simultaneously. Conceptually, individual differences could be viewed as the result of many influences. Even more recently, statistical models have been available to track changes in these many factors over developmental time.

Psychologists are in an era of catch-up. Our methodologies are catching up to our theories, and psychological models are catching up to biological ones. Biology has surprised us with its rapid move from reductionism to complexity. Medical science's devotion to theories in which all processes and individual differences are determined by underlying genes has been usurped by the new molecular biology in which all processes and individual differences are probabilistic outcomes of dynamic interacting systems including genes. The major advantage of biological scientists is that they can observe what psychologists can only infer. Compare the answer to the question, "Why did that mother not smile at her baby?" with the answer to the question, "Why did that gene turn off?" Psychologists could infer that the mother was tired from a day of hard work, was distracted by her other children, had a mother who had not smiled much at her, had a belief that the baby would not notice anyway, or had negative feelings toward the child. Biologists observe that the gene was deactivated by a methyl molecule.

Luckily, our powers of deduction have been advancing. To make an inference, one must have some possibilities in mind for explaining developmental outcomes. When one is restricted to the idea that what a person does has nothing to do with what his or her partner is doing, then the only conclusion that can be reached is that the explanation must be a personal one. The transactional model was a major step in expanding the kinds of inferences available. Maybe the explanation had something to do with the partner, the need to understand individuals in their contexts.

Because the transactional model is a metaphor, even the most devoted theorist has faced challenges in operationalizing its precepts to answer specific developmental questions. The "drunk searching for his keys under the lamplight" analogy has always been a constraining influence when researchers only view their data through the statistical methodologies they were taught and can only publish their data using methodologies with which journal reviewers are familiar. New statistical advances, many described in this book, can capture more and more of the complexity of developmental processes resulting from the interplay of personal and contextual dynamics.

Having a new model does not automatically lead to its being used in research. Because of the paucity of research explicitly examining transactional processes, I proactively asked a number of scholars to take a transactional approach to their work. We organized a conference at the University of Michigan, and the ticket of admission was having done transactional research. We assembled a stellar group of colleagues who wanted to share in this enterprise, not only presenting their own work but also learning how others approached similar problems. In the tradition of "there's no such thing as a free lunch," they all agreed to share their thoughts in this book.

The conference was one of a series primarily supported by a Behavioral Science Research Center grant awarded by the National Institutes of Health to foster interdisciplinary longitudinal research on mental health problems. Each previous conference had used an ecological frame that integrated a variety of social contexts for viewing a specific developmental issue. The previous three multidisciplinary conferences had dealt with infant mental health,[2] the transition from adolescence to adulthood,[3] and the development of behavior regulation.[4]

This book follows our overarching theme of multicontextual influences on development, with the specific goal of using the transactional model as a frame. The book is divided into four sections that dialectically move in a spiral from a description of transactional theory to how it has influenced research practice to how that practice has reciprocally led to conceptual and statistical advances of the theory. It moves from studies of the parent–child microsystem to transactions with broader social contexts of school and culture to illustrate the ubiquity of bidirectional influences throughout the social ecology, and in the end it connects transactional thinking to other dynamic systems theories, reinforcing and generalizing the systems metaphor that is at the forefront of contemporary developmental science.

The word *transaction* was in our vocabulary long before the transactional model and will continue to be long after the transactional model has been absorbed into a more comprehensive dynamic general systems theory for explaining development. But at the present time, I hope this volume provides a state-of-the-art view of transactional theory and practice that will clarify the model, offer research strategies for implementing it, and provide nodal points for others to expand on both.

Finally, I want to thank those who provided our conference with financial and logistical support at the University of Michigan, including the Center for

[2]Sameroff, A. J., McDonough, S. C., & Rosenblum, K. L. (Eds.). (2004). *Treating early relationship problems: Infant, parent, and interaction therapies*. New York: Guilford Press.
[3]Schulenberg, J. E., Sameroff, A. J., & Cicchetti, D. (Eds.). (2004). Transition from adolescence to adulthood [Special issue]. *Development and Psychopathology*, 16(4).
[4]Olson, S. L., & Sameroff, A. J. (Eds.). (2009). *Biopsychosocial regulatory processes in the development of childhood behavior problems: Biological, behavioral, and social– ecological interactions*. New York: Cambridge University Press.

Human Growth and Development, the Psychology Department, the School of Social Work, and the Center for Development and Mental Health funded by the National Institutes of Health Behavioral Science Research Center. In addition to these units, a number of individuals were especially helpful and supportive, including Susan McDonough, Sherri Olson, Katherine Rosenblum, Michael MacKenzie, Erika Lunkenheimer, Suzanne Hansknecht, and Cindy Overmyer.

I

INTRODUCTION

1
THE TRANSACTIONAL MODEL

ARNOLD SAMEROFF

Transactions are omnipresent. Everyone in the universe is affecting another or is being affected by another. Everything in the universe is affecting something else or is being affected by something else. Everything is in a relationship, from the most complex society to the most elementary particle. Although much of the history of science has been devoted to discovering the separate things in the world, the results have been quite the opposite. Most important discoveries were of the relationships in the world.

Science has been devoted to two parallel goals: an empirical task of identifying ever more fundamental things and a theoretical task of understanding their "thingness." From Aristotle's earth, air, fire and water, to Mendeleyev's periodical table of the elements, to the quarks and leptons of the contemporary standard physical model, the identification of fundamental particles defined as containing nothing smaller has moved to increasingly infinitesimal units. Since the Greeks, there has been a Platonic idealization that these newer, ever smaller elements represented things in themselves, independent of their context, and that simple combinations of these fundamental units would explain everything more complex. Unfortunately for this reductionist point of view, the properties of these particles were only evidenced when they were in a relationship with other particles, when they were affecting and being affected by other elements in the system. Quarks only become meaningful in relationships with other quarks. The transactional understanding of people is only a more complex example of the transactional understanding of quarks.

A central impetus for science has always been explanation, with the ultimate hope of control through targeted interventions. The central requirement for reaching such goals is successful prediction. The decision whether to focus on things in themselves or their relationships for these predictions is based on a belief that one emphasis would be more productive than the other. If researchers' goal is to predict how children will turn out, should they focus on more basic understandings of child characteristics, more basic understandings of experience characteristics, or more basic understanding of their relation to each other? The answer to this question has varied over time before reaching the contemporary conclusion that all are necessary.

Before transactions were interactions, and before interactions were radical nativism and nurturism. The history of developmental psychology has been characterized by pendulum swings between a majority opinion that the determinants of an individual's behavior could be found in his or her irreducible fundamental units or in his or her irreducible fundamental experiences. Galton's belief in the late 1800s that all was heredity was followed by Watson's belief in the early 1900s that all was training. These themes are carried forward in contemporary behavior genetics and learning theories. Yet in the midst of their hyperbole both Galton and Watson were interactionists. Galton (1876) acknowledged that inherited influences would be most expressed by individuals when differences in nurture did not exceed others of similar social rank in the same country, and Watson (1914) acknowledged that the effectiveness of habit training would be facilitated by knowledge of an animal's individual instinctive responses.

The interaction of nature and nurture is the current mantra of developmental science for explaining the progress of children from birth onward. On the biological side is the construct of range of reaction in which a genetic influence can be additively associated with a diversity of outcomes depending on environmental circumstance. This position has been sharpened by Turkheimer, Haley, Waldron, D'Onofrio, and Gottesman's (2003) demonstration that estimates of hereditability increase as the resources in the environment increase. In conditions of poverty with few resources, heritability is quite low, but it becomes much higher as income, social circumstance, and educational opportunities increase. In other words, if genetic variation is to express itself, it needs an environment that supports that variation. The inherent sensory and motor capacities that influence virtuoso violin playing would never be evidenced in a society where there were no violins. On the experience side are acknowledgements of individual differences. Slow versus fast learners, more attentive versus less attentive students, and more anxious versus less anxious children will all respond differently to the same educational or social experiences. Experience is still considered the driving force of development, but experience is facilitated or impeded by differences in intelligence or personality.

From a predictive perspective the interactionist strategy of including measures of both the individual and the environment is a better strategy than using either alone. This is evidenced by a myriad of studies on the consequences of such biological differences as perinatal complications after which supportive experiences bring affected children into the normal range or of adverse environments in which resilient children are able to offset abusive experiences. Despite the predictive practicality of the interactionist model, it remains theoretically flawed. At a fundamental level it is rooted in the premodern Newtonian tradition in which objects are not changed by their experience with other objects. The interactional model has not advanced to the relativistic quantum models of contemporary physics in which every relationship changes the behavior, if not the nature, of the object. Nor does it reflect the current biological revolution in which every step in the life process is a transaction. The genome was formerly believed to contain the blueprint for life; now it is seen to be a set of possibilities rather than determinants for a variety of life forms. Every cell has the identical genome, yet each is different as a result of the differences in developmental experience in different relationships to other biological activities.

The first set of developmental relationships is now conceptualized as an epigenome containing the genes plus regulators, such as methyl groups. Depending on past or current biological experience, such as whether the genes come from the mother or father, and even psychological factors, a different set of methyl groups will allow the expression of some genes but not of others. The new excitement in the study of gene–environment interactions is rooted in demonstrations that the same genes can move children on the path to have more or less aggression depending on the quality of their parental care (cf. Caspi et al., 2002). Methylation changes the expression of the genome, and the expression of the genome changes the methylation.

The original transactional model used to explain human development was a more general theoretical statement about the relation between any living entity and its experience, with epigenesis as a key example. It was derived from a dialectical philosophical tradition and overlaid on emerging empirical data about the relation between nature and nurture. Conceptually, it was in tune with the organismic model of development (Overton & Reese, 1973) that emphasized the mutually constitutive dynamic interplay between individual and context and was in accord with the principles of contemporary general systems theory (Sameroff, 1983). Direct developmental predictions from characteristics of children or their environments taken alone were not efficient. Predictions from either linear or nonlinear combinations of child and environmental characteristics were more efficient but were still based on the premise that children and their environments could be treated as constants over time. The ultimate utility of the transactional model will be in the improvement of developmental predictions.

In the transactional model, development of any process in the individual is influenced by interplay with processes in the individual's context over time. The development of the child is a product of the continuous dynamic interactions of the child and the experience provided by his or her social settings. What is core to the transactional model is the analytic emphasis placed on the bidirectional, interdependent effects of the child and environment.

The central question for this book is whether transactional research designs are producing more information than interactional ones. How closely can one get empirical studies to fit theoretical premises? In an ultimate sense, uncertainty principles proclaim that such a fit is impossible, but science is based on the belief that researchers have not yet reached the limits of the possible. Occam's razor as a principle and epidemiology as a methodology are directed at finding the simplest predictive formulations for any outcome. However, as factors in more and more settings are found to be relevant to individual development, the lowest common predictive equation becomes more and more complex. Still, methodological and statistical advances can move researchers much further along toward understanding the complexity of development without worrying yet about theoretical limitations.

INTERACTIONS AND TRANSACTIONS

Sameroff and Chandler (1975) tried to answer a difficult empirical question. Why did infants with a variety of medical anomalies, such as preterm birth or anoxia, not grow up to have the expected cognitive and emotional difficulties? The then-current medical "main-effect" model predicted a linear connection between biological problems and psychological problems (still a theme). Unexpectedly, the vast majority of at-risk children grew up well within the normal range of developmental functioning, as did a sizable minority of children with frank physical disorders. Where did the problems go?

The Sameroff and Chandler insight was to ask a similar question about deviant parenting. The child abuse literature was making a strong argument that maltreating parents had a distinct personality that did not bode well for their offspring, another linear, main-effect prediction. Again there was an unexpected finding that parents with personalities that were associated with neglecting and abusive parenting styles had many children who were not abused or neglected. How could these at-risk parents have well-functioning children?

The Sameroff and Chandler proposal was that in the first case, high-resource environments, as found in higher socioeconomic status groups, helped the child compensate for earlier difficulties, and in the second case, the child that was abused was reported to have temperamental difficulties that triggered parental maltreatment, in contrast to his or her siblings who were easier to deal with. These developmental effects could be explained with an interactional model. Good parenting could compensate for infant problems, and well-behaved children could offset the abusing tendencies of their parents, although more detailed analysis would be required to determine whether these were linear additive or nonlinear protective effects. Did a high-resource environment foster development in all children as a *promotive* effect, or did it have its greatest positive influence on high-risk children, the more elusive *protective* effect (Sameroff, 2000)? Did an infant with a good temperament have a more positive developmental course within a wide range of parenting practices, or did good temperament have its greatest effect with parents who were prone to be abusive?

The interactional model substantially increased the general efficiency of predictions leading from risk factors to poor developmental outcome; however, it was limited both methodologically and theoretically. The advances in statistics of the 1970s were beginning to solve some of the analytic problems in a move from predominantly analyses of variance to regression analyses in which interactions could be more readily tested. However, those statistical advances could not answer two questions central to the problem of development: What were the processes or etiological mechanisms that led from risk to disorder, and how could the relationship between children and parents be described?

The underlying problem in the interactional model was that child and environmental variables were defined as static entities, whereas in reality they are dynamic, both in internal and external processes. Neither constitution nor environment is necessarily constant over time. Developmental science has the opportunity to describe lifelong changes in the individual, and with the addition of life course perspectives, increasingly to describe changes over time in the environment as well. The use of growth curve statistical analyses is permitting numerical views of these developmental changes. However, these methods are only beginning to touch on the more complex issue in development, how to measure developmental changes in children and the environmental settings in which they live that are interdependent and change as a function of their mutual influence on one another.

A simple example of the difference between a transaction and interaction between two individuals may be of use here. Two individuals meet on the street, talk with each other, and move on. In the *interactional* sequence the dialogue is something like

A: Hello.

B: How are you?

A: I'm fine.

B: That's nice.

From the outside it may look like a transaction in that each individual's response is contingent on the response of the other, but the key question is, Were they changed by this social episode? For an interaction the answer is no. This is a repeated social experience that contains no novelty or adaptive pressure. Each has some well-practiced schema for such a meeting that will continue to have stability after the meeting. For a transactional sequence the schema has to change, adapt, develop, which requires something new in the experience. Such novelty would occur if the dialogue was

A: Hello.

B: How are you?

A: I'm dying of cancer [*instead of the expected "I'm fine"*].

B: Whoops!

If the "whoops" is the result of a new appraisal producing, for example, a feeling of empathy followed by a sympathetic response, then this is a transaction, and B has a modified schema for future meetings with A. But in the case of an unfeeling or inattentive B, there may still be a response of "That's nice," and no transaction would occur. Such is often the case with parents or teachers who are insensitive or inattentive to the individual state and needs of the children they are socializing or educating. Many have a one-size-fits-all model that they can frequently articulate, "This is the way that I have been teaching for 20 years, and I am not going to change just because some student can't deal with it," or more poignantly, "I know that I should be more attentive and understanding to the needs of my child, but I just don't have the time or energy."

The transactional model stresses the plastic character of the environment and of the organism as an active participant in its own growth. In parallel with the cognitive revolution in psychology, it reflects a change from the stimulus–response learning theories that view the child's behavior as simple reactions to environmental contingencies to the organismic–constructivist perspective in which children are thought to be actively engaged in attempts to organize and structure their world. Children in this view are in a perpetual state of active reorganization and cannot properly be regarded as maintaining an inborn trait or habit as a static characteristic. Developmental changes are defined by changes in the way the child interacts with experience. They are driven by new complexities in either the individual or experience that require new adaptations in one or the other. In some areas the complexity already exists in the experience; in others it comes into play as the child reaches increasing levels of maturity. Language is an example of relatively stable complexity in the environment of adult speech in which the child comes to understand more and more of what is being said by others. Education is an example of increasing complexity, as each step in formal or informal tuition is keyed to the child's ability to master the previous one. In these instances if a child shows continuity of behavior, it indicates that development has stopped because of a failure in either the child or the environment to become more complex.

Normatively, this stage of equilibration between an individual and context is called *adulthood,* when education ends and employment and family roles are stabilized. At this point transactions become interactions. Individual and context are no longer in a state of adaptation. The individual has developed responses to most contextual experiences, and the context is no longer increasing in complexity. This idealized definition of adulthood may have existed in traditional cultures in which social and technological changes occurred only sporadically across many generations. In contemporary society the rapid pace of social and technological change continually adds to the complexity of experience on the one hand, and the increasing self-reflection associated with the growth of self-actualization practices exemplified by positive psychology adds to the complexity of thought on the other. The experience of the cell phone generation is quite different from that of their parents, and in the future their increased finger dexterity will probably be thought of as part of the human genetic blueprint rather than a transactional product of text messaging.

There are also equilibrations in which a stable nonnormative continuity in individual–experiential interactions could result from physical or mental limitations in the individuals or their experiences, occurring before adult functioning is reached. Developmental disabilities place restrictions on the range of contexts that can be experienced. This is most clear in sensory anomalies of blindness and deafness. Cognitive disabilities are a more complex situation, but low intelligence and autism are classic examples in which children cannot make meaning of the complexity of the environment, placing a ceiling on their functioning. Experiential disabilities also result in stability of functioning that is often interpreted as a trait. Where no schools exist, children will all be academic failures. Even where they may exist in low-resource communities, school graduates function at depressed levels of competence.

How do researchers know that these are not individual traits of incompetence? Because of the long history of successful interventions, demonstrating advances in even the most disabled child (Shonkoff & Phillips, 2000). The most obvious examples are the successful development of the perceptually handicapped. When schools have altered their curricula for the blind and deaf and provided prostheses, such as hearing aids and braille, they have enhanced the abilities of such children to engage the complexity of experience. As a result, their academic and professional achievement increasingly overlaps with the nonphysically challenged population. The case for the neurologically impaired is more complicated, but efforts to shape experience to meet the individual needs and cognitive pace of disabled children form a national educational industry. Deinstitutionalization and integration into regular classrooms give clear evidence that individuals who had been thought to have reached their maximum level of achievement reveal surprisingly advanced levels of functioning when environmental opportunities are expanded. However, where these expansions do not occur or are not matched to the child's level, transactions will not occur. In such circumstances children will not be advanced by new experiences (see chap. 9, this volume, for an example of this point). In short, consistency in individual behavior is not necessarily evidence for personal traits but rather consistency in the processes by which these traits are maintained in the interactions between child and environment.

REGULATION

Another major shortcoming of the interactional model is that it ignores the issue of regulation. There is not only change but a seeming directionality in development. In light of the great variety and range of influences on development, the number of physical and psychological outcomes is surprisingly small. Initially, this uniformity of developmental pathways was explained by some vitalistic force that more recently was thought to reside in the genes. Now it is understood to be a consequence of biological transactions. The course of evolution has been to provide mechanisms that lead to typical developmental outcomes under most circumstances. For example, when epigenetic activity moves in a direction not supported by the fetal environment, either the activity changes or the environment changes. When the limits of such adaptations are exceeded, the fetus is miscarried or aborted, a more extreme form of the regulation of child–environment transactions. Any understanding of deviancies in developmental outcome must be seen in the light of this regulative tendency that appears to move children toward normality in the face of pressure toward deviation at this early biological level. With the current blossoming of research on regulation has come a realization that similar mutual adaptations occur at the psychological level. Most of the rhetoric is about self-regulation giving the illusion that regulation is a property of the individual, but it must be emphasized that self-regulation can only occur if there is a social surround that is engaged in "other regulation" (Sameroff & Fiese, 2000). It is this regulation by others that provides the increasingly complex social, emotional, and cognitive experiences to which the child must self-regulate and the safety net when self-regulation fails. Even early functional physiological self-regulation of sleep, crying, and attention is augmented by caregiving that provides the child with regulatory experiences to help him or her quiet down on the one hand and become more attentive on the other.

The complex construct of regulation is at the core of modern developmental theory, but three amplifications are necessary. First, evolution is not confined to biological change, and the resulting transactions occur in increasingly complex human social settings and experiences. Second, within this evolutionary process, individual development remains a coconstruction of the child's self-regulatory capacities and the social world's other-regulatory capacities that facilitates or impedes this development. The third amplification is that there are some children whose self-regulatory capacities are so compromised that existing societal coregulations to enhance development are ineffectual, and there are some environments so chaotic or malevolent that no child can attain a positive outcome. As indicated previously, in these circumstances there may be no transactions because the child is unable to differentially experience the environment or the environment is unable to differentially experience the child.

IMAGING THE TRANSACTIONAL MODEL

Understanding complexity has its own developmental course, and images offer both a powerful simplification and a dynamic understanding that is less easily conveyed by words. Even Einstein regarded his imagination as a necessity to understand and explain the universe. The original reification of the transactional model was an image (see Figure 1.1) presented in an early article (Sameroff, 1975) that highlighted the reciprocal essence of the model over time. Although the Cs and Es in Figure 1.1 are labeled *constitution* and *environment*, for this discussion they can be relabeled as *child* and *experience*. By ignoring any continuity in child or experience the figure highlighted the dynamics of the relationship between them as the primary motor for developmental change. Moreover, the environment was reduced (or perhaps raised) to the same status as the constitution in that transformation rather than identity seemed to be the fundamental process.

The examples in the original formulation were devoted to proximal social-emotional processes and distal cognitive ones. In the social-emotional realm the discussion explored how young children shaped parenting behavior through individual differences in temperament. In the cognitive realm the discussion was how socioeconomic status shaped the consequences of early biological problems. If the formulation of the transactional model had used more proximal cognitive examples, it would have been seen as analogous, if not

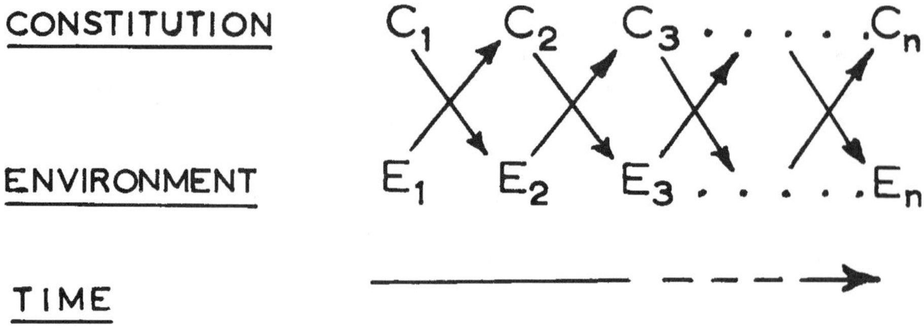

Figure 1.1. Original depiction of transactional model (Sameroff, 1975).

homologous, with the constructivistic perspective of the cognitive revolution that had begun with Piaget on the child side and Vygotsky on the experience side. From Piaget's perspective the environment has no active role in development, mainly providing the complexity that the child will come to make sense of. From Vygotsky's perspective the environment is an active force in development, in that socialization and particularly education are based on fitting experience to the developmental status of the child, his zone of proximal development. As children create their understanding of the world, the world is made more complex through steps of a curriculum to move them along toward some societal goal of mature thought. One purpose of this volume is to redress the deficiency in the cognitive side of the transactional model with several chapters devoted to effects of culture and schooling on children's thinking.

The transactional representation produced a new way of looking at things and for things. Whereas before developmental research was solely devoted to seeing the effects of things done to children, afterward research could be directed at examining how children's individual and developmental differences affected their education and socialization.

This image was very useful for the theoretical purposes of emphasizing bidirectional effects but was of less use empirically because it ignored any continuity in either child or context over time. Statistical modeling of transactional processes requires a monitoring of specific factors in the child and context over time to provide a basis for determining when these factors are changed by each other. A second figure was designed to better reflect a measurement model that could serve as a basis for testable hypotheses (see Figure 1.2). In this figure there are measures of the environment and the child across time for assessing normative developmental change as a background to the analyses of the child's effect on the environment and the environment's on the child. Patterson and Bank (1989) found the model useful to illustrate how parents' coercive childrearing behavior and children's oppositional behavior were mutually reinforcing in the developmental production of aggressive children.

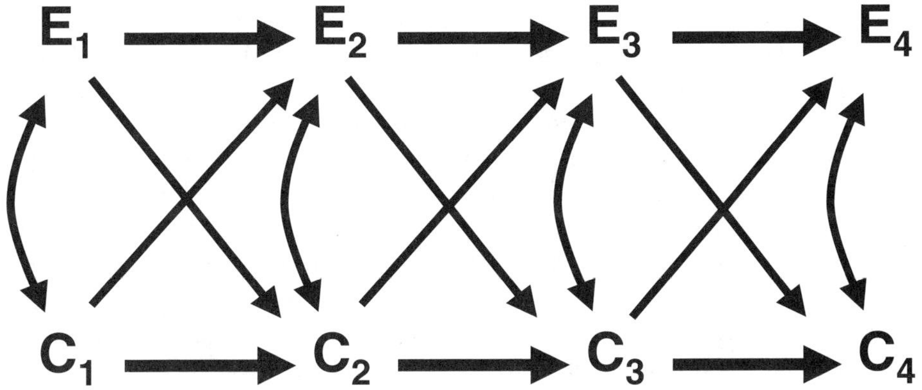

Figure 1.2. Transactional model with continuities in child and environment.

Once there is a measurement model, a new range of questions become manifest. What do the Cs and Es stand for? Usually the diagonal arrows between Cs and Es have meant observable behavior as in the Patterson and Bank example, but they also need to include the organization of that behavior or experience. Traditional research on child development has emphasized the child's use of biological capacities to gain experience and the role of experience in shaping child competencies, but there has been far less attention to how that experience is organized. Indeed, the organization of experience is explicit in the great amount of attention given to curriculum development in educational programs such as mathematics, but far less attention is given to the implicit organization of experience found in the family and social context. In the parenting domain the mother may engage in one set of physical or verbal behaviors to get the child's attention and when that is accomplished engage in another set of behaviors aimed at another goal such as feeding or preparing for bed. In the educational domain the teacher will engage in one set of behaviors to teach addition and when the student can accomplish that task move on to a second set of behaviors to teach subtraction.

In each social unit, whether parent or school or culture, there must be some level at which behavioral responses are organized, and it is that structure that is impacted when changes occur. For the individual, C representations can range from infant working models to preschool metacognitions to adolescent identities. In the simpler case in which the transaction is between two individuals, whether peers, child and parent, or two parents, then the Es are just another set of Cs. For presenting the theory it is convenient to depict both the behavior and the system that produces it within one symbol, but in the design of empirical studies these must be unpacked. When there is the opportunity to experimentally separate them, there will need to be separate constructs for the behavior and meaning system.

In chapter 2, a measurement model for transactions is described that separates behavioral and representational levels for assessing experience and the child. Is serves as a framework for empirical research on the transactional model and interpreting the research presented in the rest of this book.

Because the primary concern of the model is to understand child development, for the most part the Es represent experience provided by the context in the service of changing the child represented by the Cs. In this case experience is much more complex than that provided by one agent and represents the sum of potential responses from the context taken as a whole. In the case of parent–child encounters, the most researched parent representation is child-rearing practices, and there are a myriad of models for these from the eclectic in which the practices are sorted from a set of observations (Baumrind, 1967) to the circumplex in which parenting behaviors can be explained in a two-dimensional array (Schafer, 1959) to the hierarchical in which parenting practices are nested within parenting styles (Darling & Steinberg, 1993). When the definition of environment is restricted to parents, the affected structure can be assumed to be part of their cognitive–emotional system, however defined, and reside in the family.

Broadening the definition of the environment to include all of the social settings of the child, as in a social ecological model, requires a more comprehensive structure to include family, peer, community, and other social and cultural factors. Combining the manifold settings that affect child development required hypothesizing an overarching structure for experience that I labeled the *environtype* (Sameroff, 1989).

THE ENVIRONTYPE

The importance of identifying the different sources of regulation of human development is obvious if one is interested in developmental outcomes. Just as there is a biological organization, originally thought to be the genotype but now reconceptualized as the epigenotype, that regulates the physical outcome of each individual over time, there is a contextual organization that regulates the way human beings develop into mature members of their society (see Figure 1.3). This environtype consists of family and cultural socialization patterns analogous to the biological genotype. There will be different long-term implications if the problem is situated in the child than if it is situated in the family or in some interaction between the two. The issue is more complex from the transactional perspective in which a problem is never located completely in the child or the context but always in their relationship. As one multiplies settings, one multiplies transactional possibilities.

The child's behavior at any point in time is a product of the transactions between the phenotype, that is, the person, the environtype, that is, the source

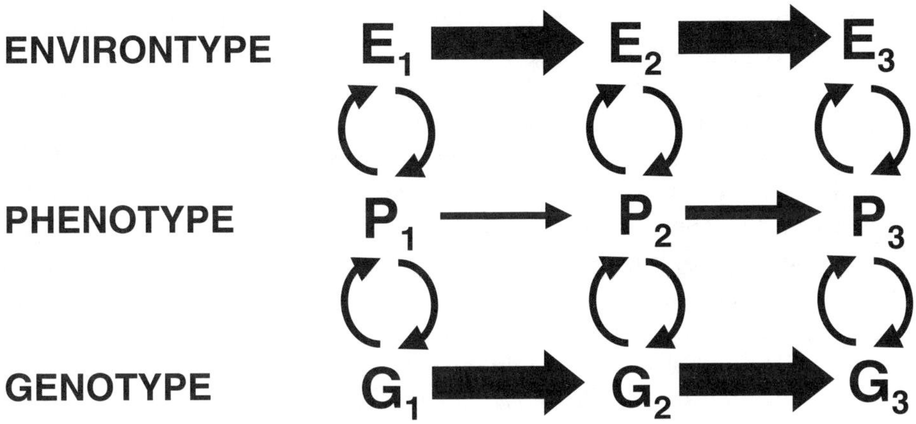

Figure 1.3. Environtype, phenotype, and genotype as mutually constitutive of individual development.

of external experience, and the genotype, that is, the source of biological organization. This regulatory system is reciprocally determined at each point in development. On the biological side the genome in each cell is identical but the particular set of genes active at any point in time is regulated by the state of the phenotype or, in the most recent formulations, the epigenome. Depending on the current chemical environment and state of methylation, certain genes are activated that alter the phenotype. The altered phenotype may then act reciprocally to deactivate the original genes and activate another set that will produce further developmental changes in the phenotype. The parallel on the environmental side is that the environtype contains a range of possible reactions to the child, but the particular regulating experiences that are active at any point in time are in response to the child's phenotype. Once the child changes as a consequence of one set of experiences, that set of experiences may be inhibited and another set activated in response to the changed status of the child.

The environtype is composed of subsystems that not only transact with the child but also transact with each other, such as the family and the school. Bronfenbrenner (1979) has provided the most detailed descriptions of environmental organizations that influence developmental processes within categories of microsystems, mesosystems, exosystems, and macrosystems. The resulting Venn diagram of overlapping settings has become a standard model for social development. Bronfenbrenner and Crouter (1983) went on to propose an empirical person-process-context model to connect the individual to settings in which transactions can be embedded in the process component. What the environtype concept adds is an organizational framework that captures the quality of the lifelong coherence of environmental influence. It emphasizes the consequence of societal reproduction and continuity as a multisetting enterprise. To this mix Super and Harkness (1986) added the physical environment in their construct of developmental niche as an important part of the world into which each child is born.

The environtype was hypothesized to be composed of a set of represented codes within such settings as the culture, the family, and the individual parent that were important for regulating development. The codes represent a set of potential responses. The general consequence of these codes is to produce an adult member of society who can play a role in reproducing that society. Such regulation during childhood is aimed at cognitive, behavioral, and social-emotional adaptation. The codes are interrelated in their evolution and in their current influence on the child. The experience of the developing child is determined partially by the beliefs, values, and personality of the parents; partially by the family's interaction patterns and transgenerational history; and partially by the socialization beliefs, controls, and supports of the culture.

Codes are different from behaviors. The environtype is no more a description of experience than the genotype is a description of the child, but it is also no less. In each case the possibilities in the codes must be actualized through processes at either the behavioral or biological level. The codes have an organizational and regulatory influence leading to adulthood but are only activated in behavior as a consequence of their interactions or transactions. The environtype can be conceptualized independently of the child; school systems can describe their curricula and some parents can describe their childrearing systems, but the existence of the environtype may not be recognized by all members of society or perhaps by any member of a particular society. It is the result of ultimate causes in the evolution of human cultures. If a culture was not successful in producing new adults to carry on roles, it would no longer exist. The proximate causes of why any particular set of parents is engaged in a particular set of childrearing behavior will have a wide range of variability as well as a wide range of coherence. For some parents there is a clear philosophy of childrearing and there are clearly desired goals for their children, such as enhanced self-direction or conformity (Kohn, 1969). For other parents there is far less reflection on their parenting, and the goal may simply be survival.

Expected, that is, normative, changes in the abilities of the developing child are major triggers for the environtype and most likely were major contributors to the evolution of each culture's developmental agenda, the implicit regulatory system that gets an individual in that culture from birth to maturity. Wachs (1991) warned against defining the environment as described by an observer when the environment as experienced by the child could be quite different. He emphasized the need for specificity in describing organism–environment interactions. As the child develops, different features of the same objective environment can come into play.

INDIVIDUAL DIFFERENCES AND DEVELOPMENTAL MILESTONES

When the transactional model was conceptualized, the focus was on individual differences in child or parent that altered developmental trajectories of one or the other. However, there are more culturally universal normative transactions that are tied to physically or socially defined milestones of the child that elicit new behaviors from their socializers (see Part II, this volume).

When researchers analyze developmental transitions, they generally make a distinction between normative ones, for example, milestones, and nonnormative ones, for example, accidents, societal changes, and historical events. This distinction is important when examining any individual life course. Although developmental milestones have usually been thought to be a property of the child, their significance is much reduced unless there is a triggered regulation from the environtype. Different parents, different families, and different cultures may be sensitive to different behaviors of the child as a regulatory trigger. For example, adolescence is defined and reacted to differently by different cultures. In some it is closely tied to biological changes; in others it precedes or follows these changes; and in others it is independent of these changes. Before the Soviet space advances of the 1950s there was little interest in infant learning capacities in the United States. Afterward infant intellectual competence, especially the ability to perceive and learn, became an important milestone for the beginning of educational efforts to catch up with the Russians.

An example of a trigger for many changes in social experience is the physical milestone of walking. From the child's perspective walking opens a world of new opportunities for experience, but from a social perspective it opens the young child to greater risk of danger and usually results in restrictions designed to protect the child. Cognitive milestones also can act as triggers to educational practices or systems. In the Vygotskian zone of proximal development, the steps in instruction are focused on moving the child from one level to the next, often through the use of cultural artifacts as elaborated by Gauvain (see chap. 9, this volume). When society bureaucratizes such instruction, it is no longer triggered by each child's developmental level but by the level of the typical child in that grade or, more detrimentally, by the level the teacher thinks is typical (see chap. 10, this volume).

CONTINUITY AS A BACKGROUND FOR TRANSACTIONAL DISCONTINUITY

The transactional model emphasizes discontinuities when parent and child mutually change their normal way of doing things. However, to recognize changes, one has to recognize continuities. Parents are far more self-regulated than children, so one would expect them to be less influenced by new experiences than their offspring. In the same vein, societies are far more formally self-regulated, for example, by means of codified legal systems, than either parents or children. However, recent history has shown a number of occasions when collective action by groups of individuals supporting the rights of women and the handicapped have had an impact on the cultural code and the legal system in the United States.

The asymmetrical power to influence another level is diagrammed as something to take into consideration when looking for transactional effects. The downward effects are larger from the societal level to the parent than from the parent to society and larger from the parent to the child than from the child to the parent. This may change developmentally as individuals move from childhood into adolescent and adulthood, when the influences from offspring to parent may be stronger than those from parent to offspring.

An interesting corollary may be operating in modern cultures that are more open to social and technologically driven change. The effect of children on their parents may be more powerful if parents are less constrained by sociocultural regulations. Instead of treating all children or all children of a certain age the same, as specified in more traditional cultures or religions, parents may become sensitive to and even seek to identify individuality in their children so that they can provide them with special experiences.

THE FUTURE OF THE TRANSACTIONAL MODEL

This chapter began with a discussion of the determining power of relationships at every level of known science. With that as a premise, researchers hoping to understand development must devote their attention to the study of relationships that change, maintain, and then change again the characteristics of the participants. Dialectical transactional models arose in attempts to understand how individuals came to know the world cognitively, especially in the relation between the knower and the known (see MacKenzie & McDonough's discussion of Dewey & Bentley's transactional theorizing in chap. 3, this volume). Here researchers have been interested in children and their socializing partners as both knowers and known, mutually constituted through their transactions.

The transactional view has become central to current models of regulation and self-regulation that are permeating the developmental literature (cf. Boekaerts, Pintrich, & Zeidner, 2000; Bradley, 2000). The individual has a major role in modifying social experience through both elicitation and selection processes and in modifying biological experience through stress management and medication (Cicchetti & Tucker, 1994). These regulations provide clear evidence for the biopsychosocial unity of human functioning.

Scientific understanding can be a foundation for controlling the fates of individuals and societies. The transactional model has been used as a basis for many intervention programs to improve developmental outcomes for children and families (Sameroff, McDonough, & Rosenblum, 2004). Characteristics of the child in the dyadic interactions between the child and parenting figures and in the child's ecological and socioeconomic circumstances are open to change. The intervener uses a transactional analysis to discover the conditions under which positive discontinuities could occur—when a change in one partner has the opportunity to reorganize the behavior in the other or a change in one setting reorganizes another setting. Such analyses identify opportunities and also set limits for intervention efforts to improve developmental success.

Transactions are easiest to describe in the relationships between parents and children, but children and their parents are involved in many ecological settings that are also changing and being changed by their participants. Explaining developmental outcomes requires attention to these multiple sources of influence as well as the parent–child dyad. This issue is clearest in intervention studies in which the interveners are part of the system, but it is equally true of all studies beyond infancy in which the parent–child relationship begins to pale in the face of peer and school involvements that occupy more of the youth's time.

As far as the transactional model is concerned, several things are clear. Children affect their environments and environments affect children. Moreover, environmental settings affect and are affected by each other. These effects change over time in response to normative and nonnormative events. Children are neither doomed nor protected by their own characteristics or the characteristics of their caregivers alone. The complexity of the transactional system opens up the possibility for many avenues of intervention to facilitate the healthy development of infants and their families. To get evidence of the multidirectional chaining of such influences requires longitudinal research that pays equal attention to the details of each individual and setting such as the research described in the following chapters.

REFERENCES

Baumrind, D. (1967). Child care practices anteceding three patterns of preschool behavior. *Genetic Psychology Monographs*, *75*, 43–88.

Boekaerts, M., Pintrich, P. R., & Zeidner, M. (Eds.). (2000). *Handbook of self-regulation*. San Diego, CA: Academic Press.

Bradley, S. J. (2000). *Affect regulation and the development of psychopathology*. New York: Guilford Press.

Bronfenbrenner, U. (1979). *The ecology of human development*. Cambridge, MA: Harvard University Press.

Bronfenbrenner, U., & Crouter, A. C. (1983). The evolution of environmental models in development research. In P. H. Mussen (Series Ed.) & W. Kessen (Vol. Ed.). *Handbook of child psychology: Vol. 1. History, theories, and methods* (4th ed., pp. 357–414). New York: Wiley.

Caspi, A., McClay, J., Moffitt, T. E., Mill, J., Martin, J., Craig, I. W., et al. (2002). Role of genotype in the cycle of violence in maltreated children. *Science, 297*, 851–853.

Cicchetti, D., & Tucker, D. (1994). Development and self-regulatory structures of the mind. *Development & Psychopathology, 6*, 533–549.

Darling, N., & Steinberg, L. (1993). Parenting style as context: An integrative model. *Psychological Bulletin, 113*, 487–496.

Galton, F. (1876). The history of twins, as a criterion for the relative powers of nature and nurture. *Journal of the Anthropological Institute, 5*, 391–406.

Kohn, M. L. (1969). *Class and conformity: A study in values*. Homewood, IL: Dorsey Press.

Overton, W., & Reese, H. (1973). Models of development: Methodological implications. In J. Nesselroade & H. Reese (Eds.), *Life-span developmental psychology: Methodological issues* (pp. 65–86). New York: Academic Press.

Patterson, G. R., & Bank, L. I. (1989). Some amplifying mechanisms for pathologic processes in families. In M. R. Gunnar & E. Thelen (Eds.), *Systems and development: The Minnesota Symposia on Child Psychology* (pp. 167–209). Hillsdale, NJ: Erlbaum.

Sameroff, A. J. (1975). Early influences on development: Fact or fancy? *Merrill-Palmer Quarterly, 21*, 267–294.

Sameroff, A. J. (1983). Developmental systems: Contexts and evolution. In P. H. Mussen (Series Ed.) & W. Kessen (Vol. Ed.), *Handbook of child psychology: Vol. 1. History, theories, and methods* (4th ed., pp. 238–294). New York: Wiley.

Sameroff, A. J. (1989). Models of developmental regulations: The environtype. In D. Cicchetti (Ed.), *Development and psychopathology* (pp. 41–68). Hillsdale, NJ: Erlbaum.

Sameroff, A. J. (2000). Ecological perspectives on developmental risk. In J. D. Osofsky & H. E. Fitzgerald (Eds.), *WAIMH handbook of infant mental health: Vol. 4. Infant mental health in groups at high risk* (pp. 1–33). New York: Wiley.

Sameroff, A. J., & Chandler, M. J. (1975). Reproductive risk and the continuum of caretaking casualty. In F. D. Horowitz, M. Hetherington, S. Scarr-Salapatek, & G. Siegel (Eds.), *Review of child development research* (Vol. 4, pp. 187–244). Chicago: The University of Chicago Press.

Sameroff, A. J., & Fiese, B. H. (2000). Transactional regulation: The developmental ecology of early intervention. In J. P. Shonkoff & S. J. Meisels (Eds.), *Early intervention: A handbook of theory, practice, and analysis* (2nd ed., pp. 135–159). New York: Cambridge University Press.

Sameroff, A. J., McDonough, S. C., & Rosenblum, K. L. (Eds.). (2004). *Treating early relationship problems: Infant, parent, and interaction therapies*. New York: Guilford Press.

Schaefer, E. S. (1959). A circumplex model for maternal behavior. *Journal of Abnormal and Social Psychology, 59*, 226–235.

Shonkoff, J. P., & Phillips, D. A. (2000). *From neurons to neighborhoods: The science of early childhood development*. Washington, DC: National Academy Press.

Super, C., & Harkness, S. (1986). The developmental niche: A conceptualization of the interface of child and culture. *International Journal of Behavioral Development, 9*, 545–569.

Turkheimer, E., Haley, A., Waldron, M., D'Onofrio, B., & Gottesman, I. I. (2003). Socioeconomic status modifies heritability of IQ in young children. *Psychological Science, 14*, 623–628.

Wachs, T. D. (1991). Environmental considerations in the study of nonextreme groups. In T. D. Wachs & R. Plomin (Eds.), *Conceptualization and measurement of organism–environment interaction* (pp. 44–67). Washington, DC: American Psychological Association.

Watson, J. D. (1914). *Behavior: An introduction to comparative psychology*. New York: Holt.

2
DESIGNS FOR TRANSACTIONAL RESEARCH

ARNOLD SAMEROFF

During the more than 3 decades since the transactional model was articulated, it has been referenced extensively in the developmental literature, and the Sameroff and Chandler (1975) article was selected as one of the 20 studies that revolutionized child psychology (Dixon, 2002). But from the outset a question frequently raised at scientific meetings was how to design research to test the model. Despite the many citations, there has been little empirical work explicitly examining reciprocal effects between child and context. All too often the model is used to emphasize the unidirectional effects of environmental risk factors on development, rather than the more complex bidirectional interplay between dynamic systems. Michael MacKenzie and I (Sameroff & MacKenzie, 2003) tackled the operationalization question by surveying the research strategies that had been used to test transactional models. To get a reasonable number of studies, we had to broaden our net to include any study that involved the analysis of bidirectional effects, whether the researchers used the word *transactional* or not. As a result, we identified a variety of experimental, quasi-experimental, and naturalistic designs that were more often implicit than explicit in their examination of transactional influences.

The experimental designs that were the strongest tests of the model used a direct manipulation of a parent's or child's behavior to see how the other would respond. In some of these studies adults were told that an ambiguous baby was a boy or girl (Seavey, Katz, & Zalk, 1975) or a premature or a full-term baby (Stern, Karraker, Sopko, & Norman, 2000), whereas in others children were trained to cooperate or be noncompliant with an adult (Bugental & Shennum, 1984). Bugental was a pioneer in this tradition and reviews much of this work in chapter 6. Other versions of experimental designs are intervention studies in which there is an explicit effort to change the behavior of one or the other social partner (Cicchetti & Hinshaw, 2002; Conduct Problems Prevention Research Group, 2002). Quasi-experimental designs occur when there is no possibility of randomization, as in the case of preterm (Poehlman & Fiese, 2001) or maltreated children (Cicchetti & Lynch, 1993). The consequences of these child differences can be traced through the effects on parents, siblings, and teachers to identify transactional components of their developmental outcomes.

The research chapters that follow can be considered a casebook in explicit transactional research. Each one takes a unique perspective on some part of the model, and the result is an enrichment of one's understanding of the whole theory. Each is an example of the growing body of research in several developmental domains that set about testing transactional models and disentangling complex bidirectional processes.

Although the transactional model originates from a strongly dialectic, organismic orientation, any operationalization requires a mechanistic measurement model. Dynamic processes are reduced to static scores that can then be entered into statistical analyses. The simplest of these is a two-way analysis of variance and a good theory. If the theory is that infants with difficult temperament who evoke negative reactions in their parents will have more negative outcomes (Thomas, Chess, & Birch, 1968), then one can dichotomize infants into difficult versus nondifficult temperament and dichotomize parents into negative reactors and nonnegative reactors and test for a statistical interaction effect on the outcome. Although this once seemed a simplistic approach, it fits well with the person-oriented methods that are becoming more common in developmental research (Magnusson & Bergman, 1990).

Theoretically and operationally, transactions need to be separated from interactions. Interactions are documented by finding dependencies in which the activity of one element is correlated with the activity of another, for example, when a smile is reciprocated by a smile, which elicits further smiling, and the correlations are stable over time. Transactions are documented when the activity of one element changes the usual activity of another, either quantitatively by increasing or decreasing the level of the usual response, or qualitatively by eliciting or initiating a new response, for example, when a smile is reciprocated by a frown, which may elicit confusion, negativity, or even increased anxious positivity. This is especially confusing when one statistical test of a *transaction* is to find a statistical *interaction*.

A problem that we identified in our review of the literature (Sameroff & MacKenzie, 2003) was a paucity of theoretically specific structural models. Additionally, there were few efforts to combine analyses of transactions in the parent–child relationship with transactions in the broader social contexts. The necessity of longitudinal studies with sufficient time points to assess reciprocal processes continues to be an issue (see Fogel's discussion in chap. 14, this volume). Such longitudinal investigations are required for identifying developmental periods when the child or the context may be most influential or most open to change.

In chapter 1, I laid out the path traversed from the initial transactional model depicted in Figure 1.1 with a minimum of arrows representing direction of effects to the all-encompassing environtype of Figure 1.3 with multiple pathways representing potentials for continuity and discontinuity. These are of value for theoretical purposes, but they do not directly address the question of operationalization. The chapters in Parts II and III of this book offer a variety of answers to that question both in terms of research design and statistical methods. These examples can be considered templates for use in extending the empirical domain when transactions can be considered from individuals to families to social institutions to cultures. The last three chapters in Parts IV and V are devoted to filling in a number of the methodological and theoretical gaps still remaining for the effective use of the transactional model.

The empirical chapters in this book represent a variety of attempts to be more explicit about examining transactional processes. Each author has a unique perspective on the transactional process of interest, often with a unique constellation of arrows. A useful generalized transactional model that can be tailored to most studies can be seen in Figure 2.1. In this generalized model of possible transactional processes, the original transactional model is schematized in the middle two rows representing the behavior of the parent (P) or the child (C). Although in most studies the Ps indicate a parent, usually the mother, they could also be any other socializing agent such as a teacher or more broadly a culture. But in the figure we have two additional rows at the top and bottom labeled *mother representations* (rM) and *child representations* (rC). For a complete transactional theory, researchers need to include levels at which behavior is organized and given meaning, and continuity is maintained.

Although one can speculate about the possibility of behavioral interactions between individuals without representations, generally for a set of behaviors to be maintained over time, there must be a representation that can be accessed for future performance. Any specific behavior is the output of some organizing system that provides meaning to that behavior. If transactions are to occur, there must be changes in the represented meaning systems. These include aspects of the family and cultural codes described in chapter 1. In what follows, the meaning systems for the parent or socializer might be a set of personal attributions, family ambitions for the child, a school curriculum, or a

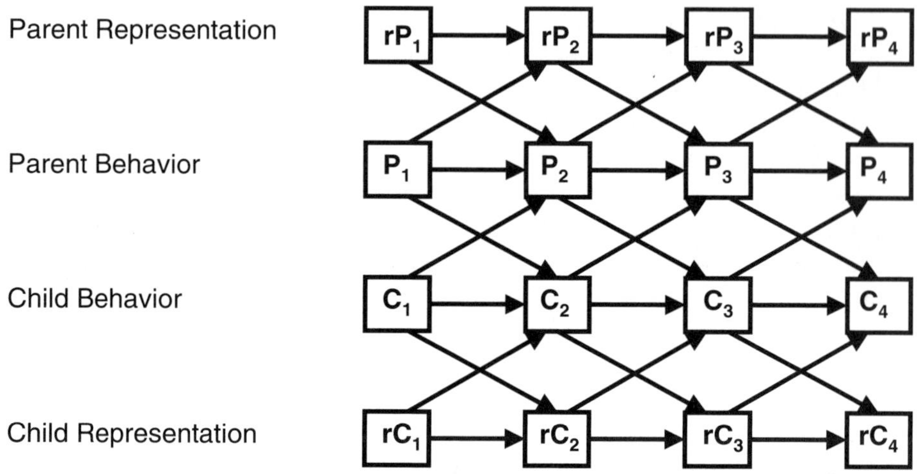

Figure 2.1. Model for operationalizing transactions among child behavior, parent or socializer behavior, child representations, and parent or socializer representations.

cultural agenda for development. On the child side, the representations include the burgeoning perceptual-cognitive understanding of the world and hypothesized working models of self and others in social development. No study presented in the book uses the full model, but each chooses some subset of constructs and arrows for empirical consideration. For most of the chapters the socializers' representations are the relevant construct, but in at least one the child's representations are involved.

How best to indicate the bidirectional influences in the model becomes an interesting question. In general, the model has been used in longitudinal research in which a measure of one member of the relationship (the parent or child) at one point in time predicts a measure of the other member of the relationship at a later point in time, usually over a duration of months or years, as indicated by the diagonal arrows in the figure. However, these transactions could be repeatedly occurring on the order of microseconds even if not measured in a particular study. This issue becomes particularly salient in chapter 11 in which Gershoff and her coauthors argue against such diagonal arrows over time and in chapter 14 in which Fogel makes the case for a very dense set of measurements across time if one wants to identify a transactional developmental change.

PARENTS AND CHILDREN

Part II of the book contains five chapters that each take a different perspective on the relations between the parent and child and between behavior and representation in the generalized transactional model. In chapter 4 by Olson and

Lunkenheimer and in chapter 5 by Shaw, Gross, and Moilanen, representations play a small role; in others they take a central role. Parent attributions are the topic of chapter 3 by MacKenzie and McDonough and chapter 6 by Bugental, and child attributions are the topic of chapter 7 by Fontaine and Dodge.

Chapter 3 by MacKenzie and McDonough is divided into two parts, a theoretical section that traces the history of transactional thinking from John Dewey to contemporary molecular biology and an empirical one that examines the transactions between parent attributions and children's crying behavior. The authors discuss how dialectical/transactional thinking is a constructive counterpoint to the linear deterministic theories that characterize the history of most sciences. In the framework provided by the generalized model, they are concerned about transactions between child behavior and mother representations, the dashed arrows between the first and third rows, excluding measures of parent behavior. Their research question is whether the later mental health consequences of excessive infant crying are better predicted by the crying or the parent's attributions about the crying. Their results are discussed in terms of a goodness-of-fit model in which a mismatch between attributions and behavior does not augur well for the child's continuing welfare.

In chapter 4, Olson and Lunkenheimer focus on the behavioral exchanges between child and parents within a regulation rubric. They integrate transactional concepts with the growing literature on behavioral regulation with a clear developmental focus. They describe how what starts out as momentary parent–infant coregulation in real time becomes more organized during the preschool period as the child's functioning becomes more complex and consistencies in behavioral regulation can be seen across situations such as home and preschool. Their primary concern in the generalized model is on transactions between the child and parent behavior levels. Increasingly as the child reaches the school years and into adolescence, the balance of who initiates relationship changes shifts from parent to child. Adding these developmental considerations is an important perspective on transactional thinking.

Shaw, Gross, and Moilanen, in chapter 5, also focus their attention on the relation between parent and child behavior. Their research question concerns the relation of parent mental health to child behavioral regulation, specifically examining the transactions between maternal depression and child externalizing behavior. In two studies spanning from toddlerhood to adolescence they direct their attention to whether maternal depression is driving the child's antisocial behavior or vice versa. Their findings support a transactional hypothesis with fascinating developmental differences in the age at which mother or child is having the greater effect. They also illustrate the complexity of identifying salient individual characteristics because different aspects of externalizing behavior, for example, noncompliance, turn out to be better predictors of parent behavior than other aspects, for example, aggression and negative emotionality.

In chapter 6, Bugental details her long history of research on parent–child transactions, which culminated in a transactionally based intervention program to reduce child maltreatment. She advances the research agenda by incorporating three levels of the generalized model into her studies: child behavior, mother behavior, and mother representations. Given a child with poor emotional regulation who is perceived as being difficult, Bugental reports different transactional processes depending on whether the mother feels threatened. In the threat case the mother's consequent negative affect leads to dysregulation of the infant and then withdrawal of the mother. However, if the mother has a benign interpretation of the child's behavior and displays positive affect, the child becomes more attentive and difficulty is reduced. On the basis of these findings, Bugental reports on a cognitively oriented intervention to increase positive responses from the mother, reducing her negative behavior.

The focus shifts from parent–child transactions to transactions between children and their representations in chapter 7; Fontaine and Dodge discuss the relation between aggressive behavior and social information processing. Using the lower two levels of the generalized model, they assess externalizing behavior and attributional biases in children from the 7th until the 12th grade. They offer an elaboration of their cognitive model of social information processing, expanding readers' understanding of the response evaluation and decision component. The Fontaine and Dodge study is among the few that demonstrate how changes in cognition can cause changes in aggressive behavior as they find bidirectional influences across that time period.

SOCIALIZATION AND EDUCATION

In Part III of this book a variety of extensions of the generalized model are described in which other levels of experience and other outcomes are examined. In the first two chapters of this section culture is added as another source of experience or as a modifier of parent–child transactions. In the next two chapters, the child outcomes shift from the social-emotional functioning that was the concern of chapters in Part II to more cognitive outcomes, especially reading.

Culture takes center stage in chapter 8 by Bornstein; he elaborates on the top three levels of the generalized transactional model—child behavior, parent behavior, and parental representations. He distinguishes bidirectional two-term transactional models in which the child affects the parent or the parent affects the child from the more complex three-term transactional models in which the effect of one partner on another redounds to change the behavior of the first. He makes the interesting point that whereas the two-term model is more common in research, the three-term model is more common in everyday life. In his discussion he differentiates between cultural representations and parent representations and gives examples of how they influence each other. He also introduces the topic of intergenerational influences in which grandparent behavior explains much of the variance in parent behavior. Bornstein presents an extensive review of how cultural constructs help to explain the relation between environments and individuals through shared meaning systems for socialization.

Gauvain, in chapter 9, continues the emphasis on the strong connection between culture and behavior but shifts her attention to cognitive development. Within a sociocultural approach she points out that social and cognitive development are mutually constituted, demonstrating the importance of transactional thinking. But the core of her transactional presentation is the description of how advanced individual cognitive abilities are products of cultural development and how in turn these abilities act to maintain and change culture. Gauvain goes on to emphasize the empirical difficulties in studying such intergenerational processes that require assessing individual and cultural change over very long periods of time.

In chapter 10, Morrison and Connor investigate the transactional relation between different teaching methods and child reading, and they report that the effect of the teacher on children in one grade can alter the effects of tutelage in the next. They focus on the behavioral levels of the generalized model but have replaced the parent with the teacher. Optimizing a child's progress requires a match between the initial skill level and the pattern of instruction. They clearly demonstrate how mismatches undermine children's achievement of literacy. In the course of the chapter they emphasize that an analysis of instructional dimensions is critical for understanding these relationships. They test their model in a randomized control trial and demonstrate that matching instruction to child creates a positive transaction resulting in better reading outcomes.

Chapter 11 by Gershoff, Aber, and Clements offers a transition to the next part of the book, on new directions for transactional research. They describe an extension of current methodologies for analyzing bidirectional effects and use as an example a longitudinal study of the relation between parent learning support and child reading ability, emphasizing the two middle behavioral rows of the generalized model. The analytic strategy of choice for transactional models had been cross-lagged panel analyses comparing the effects across large periods of time of parent on child and child on parent. Significant paths would indicate that either the parent is influencing the child or the child is influencing the parent. In a comprehensive review of lag methodologies, Gershoff and her coauthors argue that autoregressive and discrete time-lag analyses prevent the identification of ongoing transactions occurring in real time. They propose a shift to random effect continuous time-lag analyses and demonstrate how the choice of one statistical model over the other can change the direction of the sign and the interpretation of transactions between parent and child.

NEW DIRECTIONS

When the transactional model was proposed, it represented the lowest common denominator of systems theory that could connect with the then most complex level of statistical design—longitudinal cross-lag correlations. As the chapters in Parts II and III bear witness, current research on the model has become more complex both conceptually and empirically than what was available 30 years ago. The chapters in Part IV point to the future by positing both theoretical and methodological extensions that would enrich the model for solving more complex research questions.

In Chapter 12, Gonzalez catalogs the methodological issues that flow from the transactional model. The starting point is identifying multivariate approaches that can handle the many constructs involved in longitudinal analyses. Next, one must deal with analyses of multiple individuals who may be following different trajectories currently captured by latent class models. A third issue is dealing with developmental shifts that require nonlinear estimation techniques. Most important for transactions is the need for models of interdependence across individuals such as structural equation modeling. Structural equation modeling is also useful for dealing with endogeneity in which interactive systems no longer permit separation of independent and dependent variables. Finally, he deals with the problem of path dependence generally analyzed with Markov methods in which each response in a transaction between partners can constrain what is possible for the next response. Gonzalez not only elaborates on these points but also offers suggestions for the software useful for analyzing each of these issues.

The transactional model was based on dialectical principles, including the internal contradictions of the unity of opposites involved in every relationship; however, empirical difficulties have led to a major neglect of these aspects. Kuczynski and Parkin offer a major corrective to this neglect in chapter 13. They list five steps that researchers need to take to further actualize the transactional model. These include the elimination of linear thinking, conceptualizing parents and children as active agents, considering them as an active system rather then two separate individuals, seeing contradiction between partners as a primary driving force in developmental change, and searching for qualitative changes when dialectical transactional processes are occurring.

AFTERWORD

In the book's final chapter, Fogel reviews the previous chapters and discusses how successfully they interface with transactional principles. He concludes that the authors in this book are moving into new territory where transactions involve ongoing processes in which individuals are cocreated by behaving with each other. In his view, neither people nor their behaviors can be put into neat packages. The frontier for the transactional model continues to be understanding "the laws of change and the dynamics of transformation in relationship systems."

Fogel's conclusions are in accord with a contemporary reorientation in developmental science in which people cannot be understood except through their relationships. We all look forward to seeing how the dialectical contradictions between the old exclusive focus on individual characteristics and the new focus on system characteristics will be resolved and ultimately to the discovery of new questions that such a resolution will raise.

REFERENCES

Bugental, D. B., & Shennum, W. A. (1984). "Difficult" children as elicitors and targets of adult communication patterns: An attributional-behavioral transactional analysis. *Monographs of the Society for Research in Child Development, 49*, 1–18.

Cicchetti, D., & Hinshaw, S. (2002). Editorial: Prevention and intervention science: Contributions to developmental theory. *Development and Psychopathology, 14*, 667–671.

Cicchetti, D., & Lynch, M. (1993). Toward an ecological/transactional model of community violence and child maltreatment: Consequences for children's development. *Psychiatry, 56*, 96–118.

Conduct Problems Prevention Research Group. (2002). Using the fast track randomized prevention trial to test the early-starter model of the development of serious conduct problems. *Development and Psychopathology, 14*, 925–943.

Dixon, W. E. (2002). *Twenty studies that revolutionized child psychology*. Upper Saddle River, NJ: Prentice Hall.

Magnusson, D., & Bergman, L. R. (1990). A pattern approach to the study of pathways from childhood to adulthood. In L. N. Robins & M. Rutter (Eds.), *Straight and devious pathways from childhood to adulthood* (pp. 101–115). Cambridge, England: Cambridge University Press.

Poehlman, J., & Fiese, B. H. (2001). The interaction of maternal and infant vulnerabilities on developing attachment relationships. *Development and Psychopathology, 13*, 1–11.

Sameroff, A. J., & Chandler, M. J. (1975). Reproductive risk and the continuum of caretaking casualty. In F. D. Horowitz, M. Hetherington, S. Scarr-Salapatek, & G. Siegel (Eds.), *Review of child development research* (Vol. 4, pp. 187–244). Chicago: The University of Chicago Press.

Sameroff, A. J., & MacKenzie, M. J. (2003). Research strategies for capturing transactional models of development: The limits of the possible. *Development & Psychopathology, 15*, 613–640.

Seavey, C. A., Katz, P. A., & Zalk, S. R. (1975). Baby X: The effect of gender labels on adult responses to infants. *Sex Roles, 1*, 103–109.

Stern, M., Karraker, K. H., Sopko, A. M., & Norman, S. (2000). The prematurity stereotype revisited: Impact on mothers' interactions with premature and full-term infants. *Infant Mental Health Journal, 21*, 495–509.

Thomas, A., Chess, S., & Birch, H. G. (1968). *Temperament and behavior disorders in children*. New York: New York University Press.

II

PARENTS AND CHILDREN

3

TRANSACTIONS BETWEEN PERCEPTION AND REALITY: MATERNAL BELIEFS AND INFANT REGULATORY BEHAVIOR

MICHAEL J. MacKENZIE AND SUSAN C. McDONOUGH

> We are in the same position as a booking-office clerk in Paddington, who can discover, if he chooses, what proportion of travelers from that station go to Birmingham, what proportion to Exeter, and so on, but knows nothing of the individual reasons which lead to one choice in one case and another in another.
>
> —Bertrand Russell, *Religion and Science*

The study of human development has long consisted of both basic scientific efforts to expand researcher's capacity to decipher the causal laws underlying the process of individual change over time as well as a parallel applied agenda aimed at using that improved predictive power in efforts to prevent and to effect change in problematic developmental trajectories. Increasingly across scientific disciplines, efforts to find deterministic linear laws to predict the action of an individual actor—be it an atom, a fruit fly, or a human—have brought scientists to a central concept from quantum theory (Russell, 1961), namely, that all life is probabilistic. This probabilistic reality, however, does not fit well with the tendency to look for linear explanations for and solutions to an individual's or family's problems.

When infant behavior goes awry, developmental scientists seek to find causes with a degree of certainty that altering these causes would have either prevented or could now change the course of development to a happier outcome. When infants cannot seem to fully regulate their crying or sleeping or feeding, they are concerned lest this lead to later dysregulations in the emotional and social aspects of mental health. However, when the source of information is a parent, there arises the question of how accurate the information is. Is the parent perception of difficulty coming from the infant, the parents' inexperience at judging what is expected and unexpected in their children, pressures on the parents' emotional health, or a combination of the three? Developmentalists are left with a probabilistic rather than a deterministic judgment, and on the basis of that judgment, they might decide to intervene with the infant or the parent, or in the parent–child relationship.

To illuminate the transition from deterministic to probabilistic thinking, the following presentation consists of two interweaving parts. The first is critical, and the second is an attempt to be constructive. In the first part, we discuss how transactional and dialectical thinking more broadly overcome the limitations of a history of linear and deterministic approaches to the study of human development that reflect deeply rooted views about the nature of thought and behavior, especially the dichotomous treatment of subject versus object and of nature versus nurture. The basic relation between the subject and the objects in its world is inherently relational and thus dynamic. At its core, life is relational. And growth and development arise only through the irresolvable contradiction inherent in all things.

In the second part, we use the dialectical/transactional approach to explore an illustrative developmental case example focused on the relation between negative caregiver perceptions and infant crying behavior to highlight the utility of this approach for informing not just our explanatory scientific efforts but also our conceptual models of intervention and social change. A linear model would predict that the degree of infant crying should predict caregiver ratings of crying. As Sameroff and Harris (1979) pointed out, such a nondialectical approach produced a field where "we see children as being physically separate from their environments, and so we draw the obvious conclusion that they are psychologically separate as well" (p. 344). However, this does not appear to be the case. To explain this discrepancy, we advocate for an overarching dialectical approach to the study of human development in which subject and object, in this case parent and child, are seen as one unified whole.

BEYOND DETERMINISM AND REDUCTIONIST CONCEPTIONS OF DEVELOPMENT

Scientific thinking has developed gradually over a period of centuries and is more reflective of discontinuous changes in longstanding worldviews and approaches to questions involving the foundations of human thought and behavior than of an accumulation of more knowledge (Kuhn, 1970). These worldviews, or preconceptions about human development, provide the framework that guides not only the types of questions researchers ask but also where and how they look for the answers to those questions (Levins & Lewontin, 1985). The search for direct single causes of problems is not an innocuous decision. How developmental scientists define the cause of a problem is linked to their worldview and has very real implications for where their intervention efforts are focused or even the decision of whether they will intervene at all. In Western culture, with its tendencies to separate individual behavior from context (Nisbett, 2003), problems seen as having internal causes or as being a family responsibility undercut society's responsibility or role to play in resolving the problem.

Michael J. MacKenzie's journey to this current research came through earlier work in molecular genetics and a growing frustration with the reemergence of reductionist and deterministic conceptions of developmental biology that did not fit with his experience working with children reared in the foster care system. One cannot work with children who have faced such early trauma and maltreatment and not be aware of the connection between their early negative social experiences and their biological growth and development. These personal observations received additional empirical confirmation by recent work on institutionally reared children (e.g., Nelson et al., 2007). The contradiction between observations of dynamic connections between the social and biological world of foster children and the reductionist conceptual models of biological development led MacKenzie to the laboratories of Susan C. McDonough and Arnold Sameroff and their ongoing scientific efforts, which were more reflective of the bidirectional reciprocal chains of influence in early development. The work presented in the discussion that follows is reflective of that emerging intellectual partnership. Interestingly, in addition to the content produced by that partnership, the process of learning and knowledge generation itself has also been transactional in nature with an emergent focus on the reciprocal connections between basic science and intervention design and implementation.

BIOLOGICAL CONSTRICTIONS ON DEVELOPMENTAL THEORIES

Deterministic thinking in biology is characterized by the classical conceptualization of the gene, as exemplified by Watson's central dogma, in which it is defined as a coded segment of DNA that produces a polypeptide that, in turn, goes on to produce more complex proteins. What is not in this conceptualization is that this translation would never occur without a relationship with a biological context; the inert DNA molecule does not create anything in isolation (Levins & Lewontin, 1985). Each step from DNA to protein requires the activity of existing transcription machinery as well as other protein cascades and intercellular signaling. The linear prediction of what a coded base pair sequence, which is the same in every cell, will lead to is impossible without knowing its relationship to other elements in its biological environment. Even at the molecular level the story continues to become more complicated as current efforts to catalogue all proteins, exemplified by proteomics, will tell researchers little about the functionality of the protein of interest as it changes over time. A lens focused on the seemingly more fundamental genetic level, going all the way back to the point of fertilization as a starting point in hopes of predicting subsequent development in a linear manner, would still miss important transactional processes connecting the regulation of the embryo's earliest constituent factors to its later anatomy and physiology.

To adequately approach an understanding of biological development, *gene* and *environment* need to be conceptualized in a more transactional manner that places them within one unified dialectical system (Neumann-Held, 2001). This is far from a straw-man argument because reductionist and deterministic viewpoints, once thought to be soundly defeated, are threatening to reemerge in the study of human development through recent advances in genomics and the growing influence of behavioral genetics. A large part of the appeal of the inclusion of genetic analyses in researchers' studies is the recurring desire for linear causal answers even as emerging genetic evidence continues to point toward important gene by environment interactions (Scheithauer, Niebank, & Gotlieb, 2007).

Gottlieb and Halpern (2002) coined the term *analysis of variance mentality* to capture the manner in which a widely used statistical tool and overall Cartesian approach have led developmental scientists to believe that the variables that contribute to a particular outcome do in fact make independent and separate contributions to that outcome, that indeed nature can be separated from nurture. Unfortunately, the hunger for simple answers and direct causes may lead scientists to miss fundamental aspects of any system that consists of interactions in any meaningful way (Gottlieb & Halpern, 2002; Levins & Lewontin, 1985). If a particular phenomenon of interest arises from the mutual influence of a number of causative chains, and if these chains interact in even a simple manner, it is impossible to allocate an independent quantitative proportion to the causes of that phenomenon (Levins & Lewontin, 1985).

Russell (1961) cautioned scientists almost 50 years ago that there may in fact be costs associated with allowing their scientific pursuits to be guided by a practical maxim that pushes them to seek causal laws, or what he called "rules connecting events at one time with events at another" because the rules that they use have the potential to "purchase simplicity at the expense of accuracy" (p. 146). Increasingly, even in fields once thought more fundamental than developmental science—from physics to biochemistry—scientists are finding confirmation for Russell's point that early evidence from a distance suggestive of deterministic laws seems to obscure the underlying probabilistic reality and that such laws tend to be merely statistical abstractions of possible outcomes, telling scientists little of the path to be taken at the individual level.

So, which approach is most appropriate for the study of human development? The dialectical perspective takes very seriously the internal and external relations between part and whole. Parts can be parts only when there is a whole for them to be parts of. Families, for example, are made up of individuals, but can developmental scientists understand how a family works by characterizing its *parts* in detail and then looking to predict the interaction of those parts? No, they cannot, for there is no individual family member without the relationship to the family (Sameroff & Emde, 1989), the parts are only parts in relation to each other. To understand the family, developmental scientists need to understand the dynamic transformations of both the parts and the whole over time. Rather than trying to position themselves as "real scientists" by aligning with deterministic approaches, they should be patient as even the hard sciences are increasingly discovering what developmental scientists knew to be true all along. Life is built of relationships and contradictions, and it is complicated. Developmental scientists' conceptual models must be able to capture that complexity.

DIALECTICS AND THE TRANSACTIONAL MODEL

The limitation of any interactionist perspective is that it assumes two distinct and separable entities, organism and environment. But there can be no individual without an environment, and there is no environment without the individual (Levins & Lewontin, 1985). In an elegant exploration of the importance of "the named" for understanding and science, Dewey and Bentley (1949) articulated a strong definitional distinction between the terms *interaction* and *transaction*. They felt that the two terms had often been used in an interchangeable fashion without regard for the weight that the differences between them truly carry. The relationship between the knowing and the known can either be viewed in a transactional manner as one unified process or the knowings and the knowns can be separated and viewed as in interaction, such as between nature and nurture. Dewey and Bentley opted for the transactional concept as the line of thought

> where systems of description and naming are employed to deal with aspects and phases of action, without final attribution to "elements" or other presumptively detachable or independent "entities," "essences," or "realities," and without isolation of presumptively detachable relations from such detachable elements. (p. 108)

Thus, the transaction term is inherently dialectic at its core. Although words can be used to artificially divide the world, the basic reality is that they are inseparable.

On the basis of Dewey's philosophical grappling with these terms, Sameroff applied the concept of transaction to what he perceived to be a lack of appreciation for the mutually constitutive relationship between the child and the experiences provided by the context during development (Sameroff, 1975; Sameroff & Chandler, 1975). Similar dialectical conceptions had emerged in embryology, most notably in the work of Kuo (1967).

This theoretical aspect of the transactional model emerges from major theoretical streams, coupling individual and context in a relation fostering cognitive and social-emotional development. Sameroff and MacKenzie (2003) described this stream as running from philosophers like Hegel (1910) and Marx (1912) to pioneers in developmental science like Vygotsky (1962) and Piaget (1952) who emphasized the active role of the knower in creating knowledge through contradictions between knowing and the known. The dialectical core of the process is transactional in that the child is changed by experience, and experience is changed by the child's more nuanced understandings. The complex interplay of relationships of relationships in which every subject at once plays a part in many greater wholes as well as being a whole composed of dynamically interacting parts forms the basis of the internal and irresolvable contradiction inherent in all things.

In response to the strongly mechanistic and behaviorist traditions of the time, dialectical psychology emerged over 3 decades ago through the efforts of Klaus Riegel and colleagues in an attempt to contend with the continual processes of conflict and change that mark development (Riegel, 1979). Although dialectical traditions have receded from current thinking in human development, a constellation of factors is conspiring in the field and suggests a need to reexamine the place of such an orientation (see chap. 13, this volume). Dialectic thought focuses on the inseparability of subject and object and organism and environment, and on shifting the emphasis from the static to the changing and the ceaseless flux that characterizes all things (Riegel, 1979).

The dialectical notion of reciprocal influence lies at the core of transactional models of development in which mutual coconstruction is such that in one place an entity is cause and in another effect. Wherever developmental scientists begin their analysis of the parent–child relationship in development, they will find *the cause* and a series of effects that spin out from that starting point, but they cannot forget that time zero of a particular study was not the time zero. Every beginning is the result of some previous mediation and the negation of some previous negation (Dunayevskaya, 2002). Bell's (1968) reinterpretation of the direction of effects in studies of child development, drawing attention to the contributions made by the behavior and characteristics of the child, was an important first step in qualifying socialization models in which parent and culture linearly determined child behavior. The second step was realizing that new emphases on child effects did not provide a complete picture unless they were spelled out over time in a model of reciprocal influences. The transactional model is such a dynamic approach and places the ongoing parent–child transactions as central to the socialization process.

A TRANSACTIONAL CASE EXAMPLE: MATERNAL PERCEPTIONS AND INFANT CRYING

Our recent work in the area of infant crying behavior and family processes illustrates the importance of transactional models to informing both scientific and intervention efforts. The basic premise is that a child does not rear him- or herself. A child is born with little ability to regulate stressful or challenging events, relying almost entirely on the caregiver for security and regulation of infant emotional states and behavior. As the infant develops, the parent must adapt to the infant's emerging physical, cognitive, and emotional capacities. The ability of the caregiver to provide the infant with a continuing sense of organization, stability, and security that can be internalized over time into a coherent regulatory strategy is vitally important for successful development (Sameroff & Fiese, 2000).

Mothers often bring their infants to pediatricians and other health professionals because of concerns about, or frustration with, difficulties in physiological regulation, which are upsetting to the family (McDonough, 2000). The most frequently reported infant regulatory difficulties include problems involving feeding, sleeping, and excessive crying. Excessive crying problems in particular have been postulated to be associated with relationship disturbances and infant mental health problems, the precursors of later behavioral disorders and psychopathology (Lester, Boukydis, Garcia-Coll, Hole, & Peucker, 1992; Zeskind & Shingler, 1991). An area that remains the focus of debate is whether reports of regulatory difficulty in infancy represent real child illness or are projections of the parent's own issues, which may, however, create early disturbances in the caregiver–child relationship (Barr, 2000). In this light the question of where intervention efforts are best focused—either at the child's regulatory difficulty, the parent's difficulty, or at reframing the caregivers' negative perceptions of their baby—necessitates a more complete understanding of the complex processes involved (Sameroff & Fiese, 2000).

The important questions for diagnosis are the extent to which infant regulation problems are defined by problems in maternal coping capacities rather than by some quantifiable amount of crying or sleeping by the child. Because infants are seen as developing in a complex web of transactions, reciprocally influenced by individual characteristics, environmental factors, and their caregivers, there is an ongoing debate about whether concerns such as excessive crying or sleeping problems may be best conceptualized as a disorder in the infant or as reflective of a disturbance in the child–caregiver relationship (Sameroff & Emde, 1989). One theme in developmental psychopathology is the need to understand these outcomes not as residing in the individual but in the adaptiveness of the relationship between individual and context (Sameroff & MacKenzie, 2003). A requirement of such an understanding is a model of both individual and context that necessitates attention to the relationship processes involved.

These are important issues for both basic researchers attempting to understand early relationship processes and frontline professionals dealing with families who are struggling with the transition to the difficult role of parenthood. Maternal perceptions and attributions regarding child behavior are negatively influenced by the stressful effects of ecological risk and in extreme cases may play a role in the etiology of child maltreatment (Bugental et al., 2002). Sheinkopf and colleagues (2006) argued that two dyadic factors play a central role in heightening the risk for vulnerable families. The first being the extent to which mothers perceive their role as highly stressful and the second being the degree to which they perceive their infant as temperamentally difficult. Clearly these are not independent risks, in that stress and lack of support in the new parenting role increase the likelihood that the mother will perceive the infant to be difficult (Power, Gershenhorn, & Stafford, 1990).

Attempts to further developmental scientists' understanding of caregiver perceptions of infant crying and other regulatory behavior require a closer look at the role of overall caregiver cognitions in family life. An emerging parental cognition literature has highlighted multiple levels of parental belief and interpretative systems (see Bugental & Johnston, 2000; chap. 6, this volume). These range from perceptions of specific child behaviors to more general attributions and interpretative lenses to higher order representations or schemas of relationships, which have been assessed through narrative interviews such as the Adult Attachment Interview (George, Kaplan, & Main, 1985) and the Working Model of the Child Interview (WMCI; Zeanah & Benoit, 1995). Rosenblum, McDonough, Muzik, Miller, and Sameroff (2002) suggested that mothers' internal representations and working models of relationships may impact their capacity to accurately perceive and respond to their infants' emotional displays. They reported evidence that representations impact infant behavior in a pathway mediated by maternal caregiving behavior over and above the contribution of individual risk factors such as maternal depression. A next step in the theoretical model laid out by Rosenblum et al. is to examine a potential role for specific maternal perceptions of infant crying behavior as an intermediary in the link between broader representations, such as working models and caregiving behavior. Whereas Rosenblum and colleague's model went from parent's working model through caregiving behavior to infant behavior problems, ours adds the link of parent perceptions between the working model and child behavior problems.

Specifically, the purpose of our study was to determine whether (a) actual infant crying behavior or (b) maternal perceptions of the infant's regulatory behavior as problematic were more predictive of toddler behavioral problems. We also explored whether the caregiver's overall representational style of relationships moderates the link between infant crying and subsequent caregiving behavior and child behavioral outcomes. Addressing these questions has implications for the effective design and targeting of early intervention programs.

Michigan Family Study

The Michigan Family Study was designed to assess environmental risk factors and parent–child relationship disturbances to determine whether they mediate the connection between infant regulation problems and later emotional, social, and behavioral problems. Mental health during early childhood is closely associated with the quality of social and family contexts. Infants with pediatric physiological regulation problems in areas of sleep, crying, and feeding frequently have mental health problems during early childhood. The Michigan Family Study was a longitudinal study of approximately 200 families, and the study design and methodology have been described in detail elsewhere (McDonough, 2000; Rosenblum et al., 2002).

The families were evaluated for pediatric problems, environmental risk, mental health status, and aspects of parent–child relationship quality at 7, 15, and 33 months of age. Assessments included a home visit and a laboratory play session at all three waves of the study. During the home visit the mothers completed an interview and survey instrument and answered questions about the total amount of time the child cried during the day and also two questions on whether they perceived that crying to be problematic/upsetting or a bother/concern. Measures completed by the mothers included scales for maternal mental health, parenting stress, commitment to parenting, infant crying, and child behavioral regulation. For this discussion, maternal perceptions of infant crying at 7 months of age as problematic and maternal reports of time the infant actually spent crying during a typical day were related to ecological factors and scores at 33 months on the Child Behavior Checklist (CBCL; Achenbach, 1991) and the Child Behavior Questionnaire (CBQ; Rothbart, 1996).

At 33 months mother–child interactions were videotaped and scored for maternal caregiving behavior, using the 36-month coding manual of the National Institute of Child Health and Development (NICHD) Study of Early Child Care (NICHD Network, 1993). We focused on one global code of interactive behavior for *maternal supportive presence* that captured maternal positive regard and emotional support for the child.

At the 7-month laboratory visit we also used the WMCI (Zeanah & Benoit, 1995; Zeanah, Benoit, Hirshberg, Barton, & Regan, 1994), which is a systematic assessment of a parent's representation of his or her child and his or her relationship with the child. The coding and reliability procedure and use of the interview in the larger Michigan Family Study have been described previously (for additional details see Rosenblum et al., 2002). The interview focused on the parent's subjective experiences from the time of pregnancy to current interactions, beginning with a developmental history of the infant and the caregiver's relationship with the infant. The WMCI interviews were coded both for content and narrative structure and coherence and assigned to one of three representational typologies: *balanced, disengaged,* or *distorted.* Balanced refers to narratives about the infant and the caregiver's relationship with the infant that are clear and straightforward, that convey a reasonably detailed sense of the infant as a unique individual, and that demonstrate an empathic appreciation for the infant's experience. Disengaged refers to narratives that convey an emotional distance or coolness toward the infant and in which the infant is described in rather generic terms. Caregivers seem psychologically much less involved with the infant than in the balanced type. Distorted narratives are confused, inconsistent, or unrealistic.

Caregiver Perceptions of Infant Crying

Caregiver perceptions of crying as problematic or upsetting at 7 months of age were not found to be related to actual amount of concurrent infant crying. If these perceptions of infant crying were not coming directly from actual infant behavior, what factors were influencing the maternal judgments? Ecological risk factors such as parenting stress, commitment to parenting, maternal anxiety, the experience of significant life events, and family adjustment problems were all significantly associated with negative caregiver perceptions regarding infant crying and behavior, but not the amount of infant crying.

Mothers of 7-month-old babies seemed able to report negative perceptions for children who cried only minutes each day and positive perceptions of some who cried for hours. In fact, within the low-crying or high-crying half of the sample at 7 months of age, there was no difference in actual amount of crying per day between moms who viewed the crying as problematic and those who did not. The crucial finding was that by 15 months negative maternal perceptions of low-crying infants at 7 months of age was related to an increase in crying observed in these children by 15 months of age as well as behavioral difficulties at 33 months of age. No significant differences were seen in behavior problems at 33 months between early low- and high-crying infants whose mothers did not perceive their crying as problematic as measured both by the CBCL and CBQ.

In regression models, maternal perception of 7-month crying as problematic was a significant predictor of the level of crying at 15 months and of problem behavior scores on the CBCL and CBQ at 33 months. In contrast, early crying itself was not predictive of later crying or behavioral outcomes. The results demonstrate that maternal perception of 7-month-olds' crying as problematic was more important to toddler behavioral outcomes than was the actual amount of crying. Low-crying infants at 7 months of age whose mothers saw their crying as problematic went on to have worse behavioral outcomes than high-crying infants whose mothers did not see their crying as problematic.

Representations and Their Influence on Maternal Perceptions and Behavior

The data thus far demonstrated that ecologically distal risk factors were contributing to negative maternal perceptions. From Bronfenbrenner's (1979) ecological model we expected that in infancy those distal stressors and risks only impact the child to the extent that they impact more proximal systems and relationship processes. So, a logical next step was to explore potential mechanisms by which distal risk is having its impact on the child. Are these risk factors impacting caregiving interactions through their influence on the overall representations that moms have of their baby and their relationship? And are these overall representations then influencing maternal perceptions, attributions, and interpretations of specific infant behaviors in a manner that impacts subsequent maternal caregiving behavior and sensitivity?

To begin to address these questions we used the WMCI to capture maternal representations of their relationship with their infant. We examined the rates of crying being perceived as problematic or bothersome across the overall representational typologies on the WMCI—balanced, disengaged, or distorted. We found what appeared at first blush to be increased rates of negative maternal perceptions in the disengaged and distorted groups as compared with the balanced group. On closer examination, however, this main effect concealed an underlying interaction that hinted at a goodness-of-fit moderation effect (see Figure 3.1). In this graph we further break down the WMCI typologies by whether the children were in the low- or the high-crying half of our sample at 7 months of age. What we found was an interesting interaction. For balanced caregivers it did not matter whether they had a high- or low-crying infant; they were equally unlikely to be bothered by the infant's crying. But for both the disengaged and distorted typologies, the child's crying group became important. If caregivers had a low-crying infant they were not likely to be bothered, but there were significantly higher rates of negative perceptions if they had a high-crying infant. It appears that the impact of actual infant crying on parental perceptions is moderated by the security of the relationship representations of the mother. The less burdened parents are more likely to have balanced representations of their relationships and not be bothered by their child's crying. However, for the more burdened parents with unbalanced representations, an infant with higher levels of fussing and crying may be just too much to cope with.

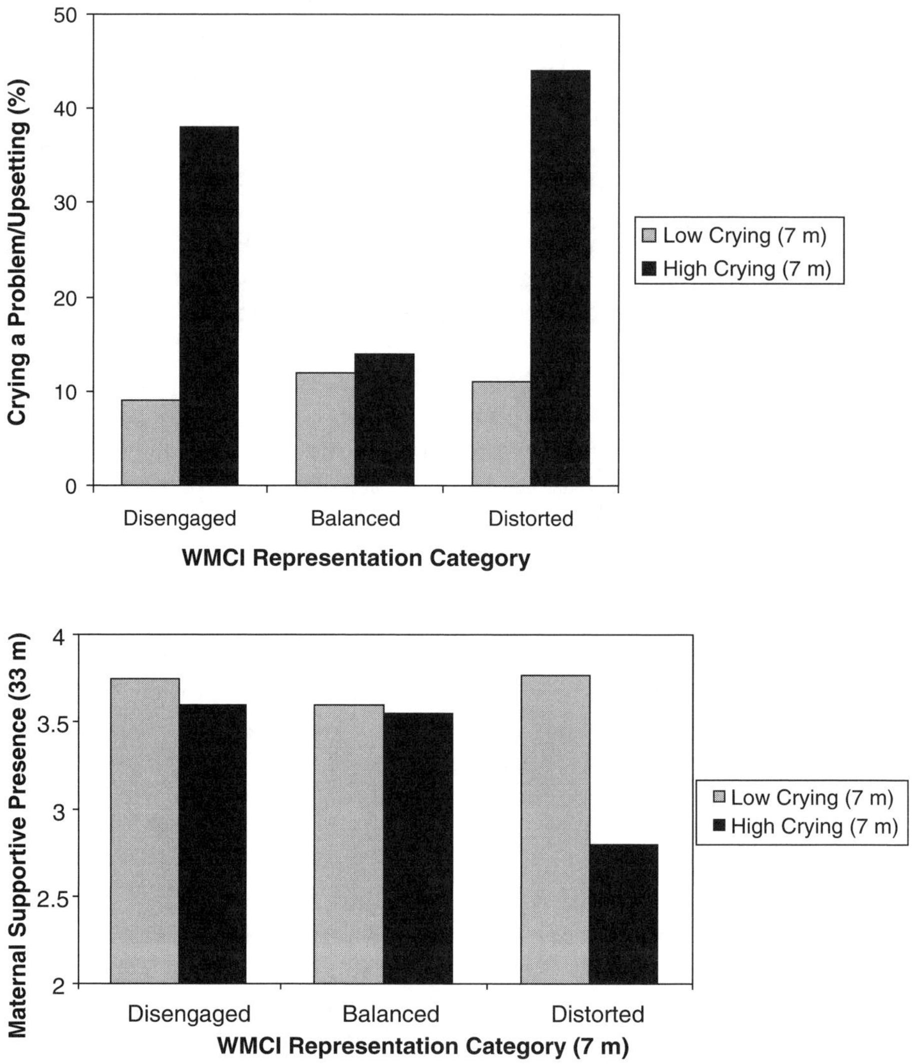

Figure 3.1. The association between maternal Working Model of the Child Interview (WMCI) representational typology at 7 months of age, caregiver perceptions at 7 months, and maternal supportive presence in a play session at 33 months of age. For mothers of low-crying infants at 7 months of age there were no differences in supportive presence in interactions at 33 months of age across representational category. For the infants in the high-crying half of the sample at 7 months, however, we found that mothers in the *distorted* category showed significantly lower levels of supportive presence with their toddlers at 33 months than for the other representational categories. m = month.

This effect involving relationship representations moderating the link between infant crying and maternal perceptions of crying behavior led us to look next at the links between maternal WMCI representational typology at 7 months and subsequent caregiving behavior at 33 months. We found a similar interaction effect based on level of infant crying behavior at 7 months for a range of maternal behaviors at 33 months. Mothers who had distorted representations at 7 months exhibited significantly lower levels of maternal supportive presence at 33 months, that is, lower levels of positive regard and emotional support, but only if their infants were in the higher crying half of our sample (see Figure 3.1).

This differential impact on maternal interactive behavior is dependent on both infant temperamental contributions and caregiver relationship representations, suggesting that the high-crying infants matched with distorted mothers may be at particular risk of interaction disturbances and subsequent behavioral difficulties. This potential mechanism led us to further explore potential links between caregiving behavior and child behavioral outcomes across representation categories. This goodness-of-fit interaction between infant crying and maternal representation typology was associated not only with maternal caregiving behavior assessed in the lab but also with subsequent child problem behavior as measured by the CBQ at 33 months of age. Toddlers of the distorted mothers received higher scores on the CBQ if they were in the high-crying half of the sample at 7 months.

Goodness of Fit as Transactional Process

Overall, we found strong support for the importance of parental cognitions to subsequent caregiving behavior and child outcomes. It appears that the level of distal stress and risk is a significant contributor to the ability of the caregiver to be emotionally available to his or her baby. This risk seems to operate at both *ecologically* and *temporally* distal levels (Bronfenbrenner, 1979) in that more macro variables from outside the microsystem of the family as well as factors from the caregiver's past history of relationships conspire to foster a milieu in which caregivers are likely to have unbalanced working models and reduced emotional capacity for the role of parenting. One potential mechanism for the impact of this distal stress on child–caregiver interaction quality seems to be through its impact on the caregiver's overall representations of relationships, which color subsequent perceptions and interpretations of specific child behavior and in so doing influence caregiving behavior.

That perceptions of infant crying are driven more by ecological risk and relationship representations than actual infant crying itself is in keeping with findings in other domains of child regulatory behavior, including feeding and sleeping difficulties (Benoit, Zeanah, & Barton, 1989; Benoit, Zeanah, Boucher, & Minde, 1992). In addition to the role of higher order relationship representations assessed by the WMCI, more specific perceptions of particular infant behavior has also been demonstrated to be impacted by the influence of external stressors on maternal functioning. Hart, Field, and Roitfarb (1999) demonstrated that mothers under the burden of stress and suffering from depression were likely to perceive their newborns as having poorer state regulation capacity following delivery than independent experimenter ratings. Unfortunately, after 1 month, experimenters rated those same babies just as negatively as the mothers did at birth, highlighting not only the impact of stress and mental health difficulty on the earliest caregiver cognitions but also the self-fulfilling processes at play whereby negative perceptions impact interactive behavior and produce a child who more closely resembles the original perceptions serving to reinforce the relational difficulty.

Our data provide additional evidence of potential mechanisms through which these caregiver representations have their impact on child emotional development. The moderating role of early child difficulty, as assessed by excessive crying, on the links between working model typology and subsequent maternal behavior in the laboratory setting at 33 months of age further highlights the complex nature of these associations. Rosenblum et al. (2002), reporting on this same sample, laid out evidence for a mediational pathway whereby the link between maternal representations at 7 months and concurrent child behavior in the still-face procedure was mediated by caregiving behavior. Sadly, the current study demonstrates that these 7-month representations continue to impact maternal and child behavior out to 33 months of age, probably as a result of stability over time both in the working models and in the disturbed relationship patterns once set in place.

The importance of maternal perceptions of infant crying to later behavioral outcomes, as compared with actual crying time, provides important insight into the transactional nature of these developmental pathways. Sameroff and Fiese (2000) have articulated a transactional model of intervention that they have called "the three Rs," which include remediation, redefinition, and reeducation. This transactional perspective can be important for the identification of therapeutic targets and appropriate strategies of intervention. The difficulties in caregiving that arise from the failure of parents to distinguish between their own emotional reactions to the child and the child's actual behavior have the potential to contribute to the etiology of later behavior problems. In this case focusing on *redefinition*, or changes in the parents' perception of the child, appears to be the most appropriate strategy for affecting change in the family system. An example of such a transactional perspective would focus on the parents' inappropriate perceptions of children's behavior working to redefine the parent's interpretation.

These are important issues for both researchers attempting to understand early relationship processes and frontline health professionals dealing with families who are struggling with the transition to the difficult role of parenthood. Maternal perceptions and attributions regarding child behavior are negatively influenced by the stressful effects of ecological risk and contribute to later child mental health difficulty and in extreme cases may also play a role in the etiology of child maltreatment (see chap. 6, this volume). Although reframing negative maternal perceptions can be an effective strategy (Bugental et al., 2002), developmental scientists must also strive to address the ecological risk factors and mental health issues, which conspire to create a situation in which negative perceptions can develop in the first place. The most effective approach may not be solely to attempt to help people better deal with their environment but also to take into account and address the external stressors and disadvantages that burden so many of today's families.

Dewey's distinction between interaction and transaction is once again relevant; conceptions of goodness of fit that assume a random coming together of independent, static characteristics of a caregiver and child that match to varying degrees fail to capture the complexity of the moderation effects we see in our data. Such static definitions also assume that all children have an equal chance of finding a good fit, yet we know that not all environments are equally open to all children and our representational data highlight that families facing higher levels of burden have a more limited range of child behavior to which they can effectively respond. Our emerging data call for more of a transactional process view of goodness of fit. We must also address the cumulative impact of risk both through initiatives aimed at buffering new parents with limited social support networks from the impact of external risks and, perhaps most importantly, by addressing the distal risks themselves. It is in this effort that we must succeed if we hope to improve the opportunity of offering a good *fit* for all babies.

CONCLUSION

In the 3 decades that have passed since the introduction of the transactional model, developmentalists have seen marked advances in their understanding of child development and attempts to bridge the disconnect between rich theoretical models and more mechanistic methodological approaches (Richters, 1997). In particular, there is some consensus that children are players in a complex web of transactions in which they are affected by the child–caregiver relationship and the surrounding environment and they are also molders of that relationship and environment through reciprocal processes (Sameroff & MacKenzie, 2003). This has necessarily led to a more complex view of development in which nature and nurture are seen as inseparable, with the social and biological worlds intertwined in a dynamic relationship in which they are continuously transacting with and changing each other. Children from families immersed in environments that present multiple social risks and disadvantages are in danger of behavioral and emotional difficulties proportional to both the level of that risk and the family's resources for coping with external stressors. The difficult task of providing a stable, structured, secure, and nurturing home for a young child is only exacerbated when parents face high levels of external stress, the presence of parental psychopathology, or a problematic history of interpersonal relationships (Sroufe, 1990).

In our empirical example focusing on maternal perceptions of infant crying behavior we found that the initial linear causal models were too simplistic, as is usually the case when we begin to elucidate complex developmental pathways. We did not find support for the idea of a *fussy baby*—qualitatively different in kind from other babies—who then turns into a problematic toddler. A more nuanced understanding came when we applied a transactional model that recognizes the reciprocal chains of influence throughout development (see Figure 3.2). Rather than seeing an inherently fussy baby leading to later prob-

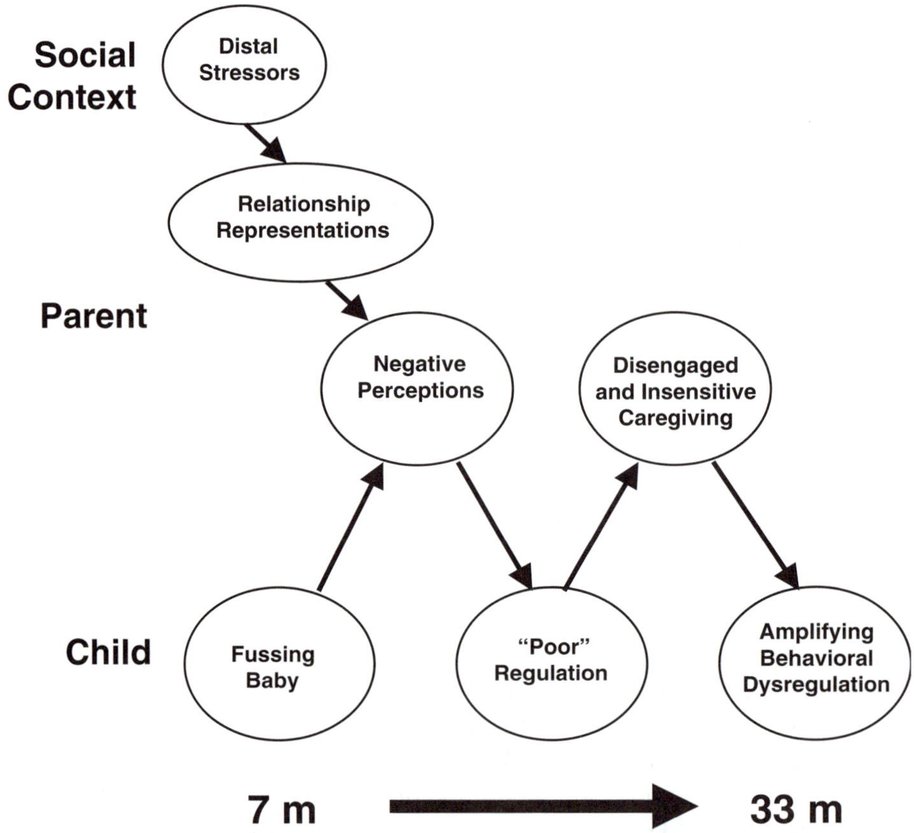

Figure 3.2. Conceptual model of the transactional processes linking negative caregiver perceptions and infant regulatory behavior problems. m = month.

lems, we saw that the spectrum of fussing behavior that babies emit is interpreted by the caregiver on the basis of his or her own expectations and goodness of fit with the baby. Negative perceptions have the potential to impact caregiver behavior, which in turn impacts the child's later behavior, serving to confirm the parent's earlier expectations and leading to further difficulties downstream. We also saw that distal risk factors in both the ecological and temporal sense, contribute to negative maternal representations and perceptions. What was often reported as an infant problem in crying and regulation seemed to more accurately reflect underlying disturbances in the parent–child relationship that played out in the caregiving interactions of everyday life.

Developmental scientists would do well to recall Russell's (1961) admonition that although a deterministic practical maxim to guide scientific endeavors may be useful in pushing forward understanding, there are potential costs that may be associated with such an approach. Developmental scientists should consider the possibility that such a deterministic approach to their questions impacts even the design and execution of their studies in a manner that may cause them to miss more dynamic developmental processes such as those highlighted by the case example in this chapter as well as others in this volume.

The gradual untangling of these complex social and biological processes will continue to be an important basic scientific endeavor, but when it comes to developmental scientists' applied work they do not need to wait for a complete resolution of these pathways before they begin to attend to potential avenues of intervention that are highlighted by these findings. In fact, developmental scientists' experiences in implementing intervention efforts are likely to point them in the direction of important avenues for future basic scientific endeavors, linking basic and applied work together in a dialectical transaction important to the process of their work as scientists and practitioners. Despite the fact that there remains much more to learn in the realm of basic developmental science, limitations in developmental scientists' knowledge base can no longer be construed as the limiting factor in their efforts to effect change in the lives of burdened families with young children.

The problem is no longer that developmental scientists lack a consensus or knowledge base for the sorts of things that need to be done. For their applied efforts, an understanding of the probabilistic transactional nature of life and relationships combined with their current knowledge is sufficient to begin to guide developmental scientists' efforts to effect change. Even stipulating the possibility that their scientific advances may never allow them to completely understand and predict with certainty the trajectory any particular child or family will take does not preclude our potential to effect meaningful change for a large spectrum of our society. If for developmental scientists' applied practice efforts they accept that there is a probabilistic array of children in any one environmental context that will have positive or negative trajectories, it forces them to focus their efforts on adapting that social context to shift the odds toward a wider proportion of children experiencing positive outcomes. These two parallel efforts, the scientific and the applied, are not mutually exclusive, but the worldview guiding both may need to be expanded to allow for both to occur. As the basic science of development continues to advance and push back the boundaries of knowledge of dynamic processes, developmental scientists' intervention and policy efforts may be best focused on the probabilistic and transactional nature of life.

REFERENCES

Achenbach, T. M. (1991). *Manual for the CBCL/4-18 and 1991 profile*. Burlington: University of Vermont.

Barr, R. G. (2000). Excessive crying. In A. Sameroff, M. Lewis, & S. Miller (Eds.), *Handbook of developmental psychopathology* (2nd ed., pp. 327–350). New York: Plenum Press.

Bell, R. Q. (1968). A reinterpretation of the direction of effects in studies of socialization. *Psychological Review, 75*, 81–95.

Benoit, D., Zeanah, C. H., & Barton, M. L. (1989). Maternal attachment disturbances in failure to thrive. *Infant Mental Health Journal, 10*, 185–202.

Benoit, D., Zeanah, C. H., Boucher, C., & Minde, K. K. (1992). Sleep disturbances in early childhood: Association with insecure maternal attachment. *Journal of the American Academy of Child and Adolescent Psychiatry, 31*, 86–93.

Bronfenbrenner, U. (1979). *The ecology of human development*. Cambridge, MA: Harvard University Press.

Bugental, D. B., Ellerson, P. C., Lin, E. K., Rainey, B., Kokotovic, A., & O'Hara, N. (2002). A cognitive approach to child abuse prevention. *Journal of Family Psychology, 16*, 243–258.

Bugental, D. B., & Johnston, C. (2000). Parental and child cognitions in the context of the family. *Annual Review of Psychology, 51*, 315–344.

Dewey, J., & Bentley, A. F. (1949). *Knowing and the known*. Boston: Beacon Press.

Dunayevskaya, R. (2002). *The power of negativity: Selected writings on the dialectic in Hegel and Marx*. New York: Lexington Books.

George, C., Kaplan, N., & Main, M. (1985). *Adult attachment interview*. Unpublished manuscript, University of California, Berkeley.

Gottlieb, G., & Halpern, C. T. (2002). A relational view of causality in normal and abnormal development. *Development and Psychopathology, 14*, 421–435.

Hart, S., Field, T., & Roitfarb, M. (1999). Depressed mothers' assessments of their neonates' behaviors. *Infant Mental Health Journal, 20*, 200–210.

Hegel, G. W. F. (1910). *The phenomenology of mind* (J. B. Baillie, Ed. & Trans.). London: Allen & Unwin.

Kuhn, T. S. (1970). *The structure of scientific revolutions.* Chicago: University of Chicago Press.

Kuo, Z.-Y. (1967). *The dynamics of behavior development.* New York: Random House.

Lester, B. M., Boukydis, C. F. Z., Garcia-Coll, C. T., Hole, W., & Peucker, M. (1992). Infantile colic: Acoustic cry characteristics, maternal perception of cry, and temperament. *Infant Behavior and Development, 15,* 15–26.

Levins, R., & Lewontin, R. (1985). *The dialectical biologist.* Cambridge, MA: Harvard University Press.

Marx, K. (1912). *Capital.* London: Glaisher.

McDonough, S. C. (2000). Interaction guidance: An approach for difficult-to-engage families. In C. H. Zeanah (Ed.), *Handbook of infant mental health* (2nd ed., pp. 485–493). New York: Guilford Press.

Nelson, C. A., Zeanah, C. H., Fox, N. A., Marshall, P. J., Smyke, A. T., & Guthrie, D. (2007). Cognitive recovery in socially deprived children: The Bucharest early intervention project. *Science, 318,* 1937–1940.

National Institute of Child Health and Development Early Child Care Research Network. (1993). *The NICHD study of early child care manual.* Bethesda, MD.

Neumann-Held, E. M. (2001). Let's talk about genes: The process molecular gene concept and its context. In S. Oyama, R. E. Griffiths, & R. D. Gray (Eds.), *Cycles of contingency: Developmental systems and evolution* (pp. 69–84). Cambridge, MA: MIT Press.

Nisbett, R. E. (2003). *The geography of thought: How Asians and Westerners think differently . . . and why.* New York: Free Press.

Piaget, J. (1952). *The origins of intelligence in children.* New York: International Universities Press.

Power, T. G., Gershenhorn, S., & Stafford, D. (1990). Maternal perceptions of infant difficultness: The influence of maternal attitudes and attributions. *Infant Behavior & Development, 13,* 421–437.

Richters, J. E. (1997). The Hubble hypothesis and the developmentalist's dilemma. *Development and Psychopathology,* 9, 193–229

Riegel, K. F. (1979). *Foundations of dialectical psychology.* New York: Academic Press.

Rosenblum, K. L., McDonough, S., Muzik, M., Miller, A., & Sameroff, A. (2002). Maternal representations of the infant: Associations with infant response to the still face. *Child Development, 73,* 999–1015.

Rothbart, M. K. (1996). *Children's behavior questionnaire: Short form Version 1.* Unpublished manuscript, University of Oregon.

Russell, B. (1961). *Religion and science* (2nd ed.). London: Oxford University Press.

Sameroff, A. J. (1975). Early influences on development: Fact or fancy? *Merrill-Palmer Quarterly, 21,* 267–294.

Sameroff, A. J., & Chandler, M. J. (1975). Reproductive risk and the continuum of caretaking casualty. In F. D. Horowitz, M. Hetherington, S. Scarr-Salapatek, &

G. Siegel (Eds.), *Review of child development research* (Vol. 4, pp. 187–244). Chicago: The University of Chicago Press.

Sameroff, A. J., & Emde, R. N. (Eds.). (1989). *Relationship disturbances in early childhood: A developmental approach*. New York: Basic Books.

Sameroff, A. J., & Fiese, B. H. (2000). Models of development and developmental risk. In C. H. Zeanah (Ed.), *Handbook of infant mental health* (2nd ed., pp 3–19). New York: Guilford Press.

Sameroff, A. J., & Harris, A. E. (1979). Dialectical approaches to early thought and language. In M. H. Bornstein & W. Kessen (Eds.), *Psychological development from infancy: Image to intention* (pp. 339–372). New York: Erlbaum.

Sameroff, A. J., & MacKenzie, M. J. (2003). Research strategies for capturing transactional models of development: The limits of the possible. *Development and Psychopathology, 15*, 613–640.

Scheithauer, H., Niebank, K., & Gotlieb, G. (2007). To see an elephant: Developmental science. *European Journal of Developmental Science, 1*, 6–22.

Sheinkopf, S. J., Lester, B. M., LaGasse, L. L., Seifer, R., Bauer, C. R., Shankaran, S., et al. (2006). Interactions between maternal characteristics and neonatal behavior in the prediction of parenting stress and perception of infant temperament. *Journal of Pediatric Psychology, 31*, 27–40.

Sroufe, L. A. (1990). Considering normal and abnormal together: The essence of developmental psychopathology. *Development and Psychopathology, 2*, 335–347.

Vygotsky, L. S. (1962). *Thought and language*. Oxford, England: Wiley.

Zeanah, C. H., & Benoit, D. (1995). Clinical applications of a parent perception interview in infant mental health. *Infant Psychiatry, 4*, 539–554.

Zeanah, C. H., Benoit, D., Hirshberg, L., Barton, M. L., & Regan, C. (1994). Mother's representations of their infants are concordant with infant attachment classifications. *Developmental Issues in Psychiatry and Psychology, 1*, 1–14.

Zeskind, P. S., & Shingler, E. A. (1991). Child abusers' perceptual responses to newborn infant cries varying in pitch. *Infant Behavior and Development, 14*, 335–347.

4

EXPANDING CONCEPTS OF SELF-REGULATION TO SOCIAL RELATIONSHIPS: TRANSACTIONAL PROCESSES IN THE DEVELOPMENT OF EARLY BEHAVIORAL ADJUSTMENT

SHERYL L. OLSON AND ERIKA S. LUNKENHEIMER

> Any truly transactional model must stress the plastic character of the environment and of the organism as an active participant in its own growth. . . . In this view, the constants in development are not some set of traits but rather the processes by which traits are maintained in the transactions between organism and environment. (Sameroff & Chandler, 1975b)

One measure of the power of a theory is how many new generations of scholars have incorporated the theory into their own thinking and research. Sameroff's transactional model had a profound impact on our development as psychologists. Prior to the proliferation of integrative paradigms in the mid-1980s, most prevailing models of human behavior and development were reductionistic monotheories. For example, children were portrayed as self-directing organisms or passive recipients of environmental influence. Individuals were the products of nature or nurture. Fortunately, a review article by Arnold Sameroff and Michael Chandler (1975b) entitled "Reproductive Risk and the Continuum of Caretaking Casualty" appeared on the required reading list for Sheryl Olson's preliminary examinations. The premise of the article seemed almost revolutionary for the time, yet made perfect intuitive sense. Here was an elegant way of thinking about the complexity that people observe all around them everyday: Children constantly change and are changed by their social environments, which in turn reflects many

This research was supported by National Institute of Mental Health Grant R01 57489 to Sheryl Olson. We also wish to thank the children, parents, and teachers who generously shared their time with us.

different sources of strain and support. Most important, the balance of stresses and supports to the child becomes transformed over time in ways that exacerbate or attenuate emerging adjustment problems. Thus, the parent of a toddler who is hard to manage may struggle to promote healthy self-regulation in his or her child; with adequate support resources, the chances for success are favorable. Similar behavioral challenges may overwhelm the coping resources of a parent with poor support, especially if he or she experiences multiple and chronic strains. Harsh or unsupportive parenting, in turn, is an important predictor of worsening and generalized problems for the child. Thus, one cannot understand individual development without understanding how the child affects and is affected by specific environmental contexts, which in turn are ever changing.

Now over 30 years old, the transactional model seems as timely as ever. It has continued to inspire the ways in which new generations of psychologists and allied professionals conceive of development, psychopathology, and intervention. As concrete evidence of this assertion, Erika Lunkenheimer is an early career researcher whose work strongly reflects the influence of Sameroff's transactional theory.

However, representing transactional dynamics in research has been a formidable challenge: There is a large gap between theory and empirical research that truly operationalizes the tenets of the theory (Sameroff & MacKenzie, 2003). As one way of trying to bridge this gap, we argue that concepts of self-regulation offer useful windows into the nature of transactional processes in development. In what follows, we show how concepts of self-regulation provide the framework for an intermediate-level theory linking broader constructs of transactional models to behavioral phenomenology.

This chapter is divided into two parts. In the first section we discuss different concepts of self-regulation and show how these concepts are essential for understanding the nature of transactional processes in development. In the second section, we show how individually focused concepts of emotion regulation can be expanded to summarize the regulation of dyadic relationships, especially parent–child interactions and consequences for the development of psychopathology. We argue that concepts of coregulation hold promise as ways of tapping the powerful social foundations of children's regulatory competence and as ways of assessing complex relationship dynamics that are linked to patterns of behavioral adjustment across development.

CONCEPTS OF SELF-REGULATION

As Michael Posner and Mary Rothbart (2000) have written, "Understanding self-regulation is the single most crucial goal for advancing our understanding of development and psychopathology" (p. 427). Achieving adequate levels of self-regulation underpins healthy social and emotional functioning across the life span (Charles & Carstensen, 2007; Diamond & Aspinwall, 2003). However, defining the concept has been an ongoing challenge. In this section, we outline definitional features and argue that concepts of self-regulation provide a family of conceptual tools for understanding the nature of transactional processes in development. Part of our argument hinges on paying close attention to the process-oriented details of individual lives. Thus, we begin with an example from our ongoing longitudinal study of children's early behavioral development. The setting is a family living room. A 3-year-old boy and his father have been asked to play together and to solve some difficult puzzles. Now, the visitor asks the father to persuade his son to clean up the toys, without doing the job for him:

Father: Can you put the toys away for me, please?

Child: NO!

Father: Please?

Child: NO!

Father: Are you playing with the baby? Can you put the baby away?

Child: I don't wanna! NOOOOOOO!

Father: You can look at 'em. But now you need to put 'em away because the lady needs to go pretty soon.

Child: NO!

Father: Well, what are we gonna do? The lady asked us to put them away.

Child: NOOO! I don't wanna put 'em away!

Compliance behavior is seen by many developmental theorists as an early form of self-regulatory competence. Clearly, this child demonstrated strong and consistent noncompliance as well as overt defiance vis-à-vis his father. But what is the significance of his behavior? Does it reflect a failure in emerging self-regulation that portends future difficulties? Conversely, the boy's behavior could reflect a courageous assertion of autonomy that may augur future leadership skills. A third possibility is that his child's noncompliant behavior reflects a *perturbation* (stressful challenge) that is fairly normal in toddler-age children.

In fact, most difficult toddlers do not develop persistent problems (Campbell, 2002; Kopp, 1982; Tremblay, 2000). Thus, we must pose an inherently transactional question: What are the developmental processes linking early vulnerabilities in the child to stable patterns of maladjustment?

Many forms of psychopathology, particularly the disruptive disorders, reflect failures in the development of normal self-regulatory competence (e.g., Campbell, Shaw, & Gilliom, 2000; Keenan, 2000; Olson, Sameroff, Kerr, Lopez, & Wellman, 2005). What do we mean by normal self-regulatory competence? Current definitions stress that self-regulation is an ongoing process of modulating attentional, behavioral, and emotional responding in ways that potentiate socially adaptive outcomes (Kopp, 1982, 1989; Posner & Rothbart, 2000; Thompson, 1994). We propose that three components are essential for understanding the nature of self-regulation: (a) integration across multilevel systems, (b) dynamic microprocesses that characterize an individual's responses to stressful challenges, and (c) the meaning of these processes across development.

Integration Across Multilevel Systems

The concept of self-regulation subsumes multiple coordinated component systems that reflect the influences of a broad range of internal and environmental factors (Cole, Martin, & Dennis, 2004; Fox & Calkins, 2003). For example, even simple acts of compliance reflect the coordination of attentional, behavioral, and emotional responses to challenging situations. These responses, in turn, are influenced by a broad range of internal traits, such as the child's temperament characteristics, and by salient state conditions such as fatigue or illness. Each internal influence is linked to biological control systems that regulate and are regulated by environmental stress (Lopez, Vazquez, & Olson, 2004; Posner & Rothbart, 2000; Rothbart, Sheese, & Posner, 2007). For example, children of depressed parents have been found to show poorer heart rate recovery after a disappointment than controls (Forbes, Fox, Cohn, Galles, & Kovacs, 2006). These children also showed an association between greater relative left frontal electroencephalogram activity and comorbid externalizing and internalizing problem behaviors. This work is just one example of how internal (e.g., physiology) and environmental (e.g., parental depression) factors might interact to influence both moment-to-moment regulatory responses and long-term regulatory capacities for the child.

Perhaps most important, children's regulatory behaviors both change and are constantly changed by environmental influences. Several lines of research have shown that self-regulation skills reflect reciprocal and transformational influences between the child and his or her social partners. For example, aggressive, noncompliant behavior tends to elicit hostile, controlling responses from caregivers, siblings, and peers (e.g., "upper limit controls"; Bell,

1971), often creating positive or *deviance amplifying* feedback cycles (Patterson & Bank, 1989). This was illustrated in a short-term longitudinal study of aggressive preschool boys and their peers, followed through a year of preschool life (Olson, 1992). In the early fall, highly aggressive, disruptive children were readily identified by teachers and by peers (e.g., "Who hits, fights a lot? Who is mean to others?"), and children's deviance status remained stable across time. Detailed behavioral observations of peer interaction sequences revealed that in the early months of preschool, peers did not respond aggressively to these children's frequent aggressive provocations. However, over the course of the year there was a dramatic escalation in the extent to which peers reciprocated aggressive behaviors from high-risk boys. Moreover, when these transactions were examined qualitatively, it was clear that aggressive boys who tried to back off from conflict were pulled right back into the fray by highly aggressive peer provocations—in other words, the victimizers became victimized (Olson, 1992; Miller & Olson, 2000). For example, Carl, age 5, was rated by teachers as highly aggressive and by peers as highly aggressive and disliked. The following sequences describe Carl's peer interactions during the final observation period (late spring of the preschool year):

> Carl initiated interaction by driving his toy spaceship into another boy's block tower. The other boy pushed his spaceship away. Carl played by himself for 1 minute. During the next seven sequences, his spaceship was smashed by different peers. Each time, Carl protested angrily and tried to hit their toys in return. Next, he knocked down a block tower that another girl and boy were building.
>
> The boy replied, "Don't! That's mean, Carl!" At this point a second boy rammed a car into Carl's ship and threatened, "You! Next time you build something, I'll wreck it!" Next, a second girl grabbed Carl's robot while his gaze was averted and taunted him by saying, "I didn't take your robot" while simultaneously waving it in his face. Carl replied, shouting angrily, "GIVE ME IT!" He reached for the robot, and threatened her by balling up his fist and saying, quietly, "I'll kick you; I'll smack you in the . . ." In response, a boy bystander slapped him and threatened "If you smack her, I'm gonna kick you." He then punched Carl five times and hit his knee forcefully with a toy car, while the girl chanted, "Ha, ha, ha, ha!"
>
> When Carl tried to move away, the boy taunted him by yelling, "Chicken, chicken!" Carl retaliated by throwing a block at the boy, then retreated and began playing alone with his spaceship. However, the girl who was involved in the previous skirmish threw a block at his ship, then grabbed his robot again. Carl grabbed it back, and pushed her. Following several rounds of mutual block throwing, the girl told another boy that

Carl knocked down his block tower (he did not do this). The boy punched him in the back, and Carl began to cry.[1]

In sum, concepts of self-regulation subsume multiple coordinated response systems within the child, such as neurophysiological, attentional, emotional, and behavioral systems. In addition, it is clear that the nature of children's regulatory responses reflect social interaction processes in various contexts that play a major role in shaping the quality of children's regulatory competence over time (see also Volling, Kolak, & Blandon, in press). Thus, what psychologists typically call *self*-regulation more accurately reflects emergent social regulatory processes at the nexus between the individual and his or her social partners. Furthermore, as shown in the discussion that follows, to understand the nature of these transformational phenomena, psychologists must consider self-regulation as a pattern of progressive responses to challenges in real time (seconds or minutes) as well as how the meaning of regulatory competence reorganizes across developmental time (months or years).

Dynamic Processes in Real Time

In some studies, self-regulation has been defined as a kind of static outcome variable. However, although outcomes at any given point in time are useful indexes of adjustment, self-regulation is an inherently dynamic construct. Too little attention has been placed on mapping components of individual children's responses to challenging events as they fluctuate in real time. Even within a single-response system, there is remarkable individual variability in processes underlying children's responses to stress. Unless studies are designed to highlight these processes, important information about individual patterns of variability is lost. For example, Lopez (2006) densely sampled neuroendrocrine markers of 7-year-old children's responses to discrete stressors administered in a laboratory setting. Children's responses to single stressors could be described as complex *chains* of neurocognitive processes that modulated the activation, intensity, duration, and expression of arousal. Individual differences in reactivity (the slope and pattern of cortisol production following stress exposure) and regulation (the slope and pattern of cortisol production following peak or maximal responding to a stressor) were extremely salient. Individual children were asked to complete a challenging task in the laboratory (retrieving an attractive prize in a Plexiglas container with the lid glued shut). Stress exposure was preceded by 30 minutes of quiet, nonarousing activity. Salivary cortisol production was assessed every 5 minutes beginning 30 minutes before the challenge through 60 minutes posttask. The baseline level, slope, time of

---

[1]Adapted with kind permission from Springer Science+Business Media, "Development of Conduct Problems and Peer Rejection in Preschool Children: A Social Systems Analysis," by S. L. Olson, 1992, *Journal of Abnormal Child Psychology*, *20*(3), p. 346. Copyright 1992 by Springer Netherlands.

peak response, and pattern of responding varied widely between individual children. For example, some showed minimal reactivity to challenge, whereas others had cortisol activation curves that resembled mountain peaks. These patterns of variability are not represented in measures that summarize average levels of responding among children in a study (Lopez-Duran, Olson, Felt, & Vazquez, in press). Keenan, Jacob, Grace, and Gunthorpe (in press) described similar levels of variability in newborns' response to routine but stressful medical procedures (e.g., a heel stick). For example, healthy neonates revealed high levels of individual variability in the patterning of their neuroendrocrine responses to the same stressor: Some showed increased cortisol responding; some showed minimal changes in cortisol reactivity; and others showed decreased cortisol levels following the heel stick procedure.

Thus, the *patterning* of the individual's moment-to-moment responses to stressful challenges may be an important precursor of children's regulatory competence across development (Keenan & Jacob, in press). However, judgments of competent versus compromised functioning must be anchored to knowledge of typical responding for children of a given age. In the following section, we briefly consider normal ranges of regulatory competence and how they change across development.

Changing Contexts in Developmental Time

Children's regulatory skills emerge across distinct developmental phases, paralleling growth in cognitive functioning (especially maturation of the prefrontal cortex; Kopp, 1989; Posner & Rothbart, 2000; Rueda, Posner, & Rothbart, 2005). Each phase is marked by different regulatory challenges that must be mastered for smooth transition to the next phase. Thus, each phase can be seen as a turning point for successful versus unsuccessful behavioral adjustment. With increasing maturation, the role and influence of the social environment changes dramatically. For example, in early infancy, control takes the form of modulation of physiological arousal states. Although infants enter the world with different styles of emotional reactivity, they are completely dependent on their caregivers. Toddlers are capable of complying with simple requests, but control remains limited to the immediate stimulus environment. During the preschool period, extremely rapid growth in self-regulation reflects more complex cognitive processes that come on line, permitting increased levels of behavioral flexibility (Calkins, in press). For example, preschool-age children can flexibly adapt to life situations that have different standards of conduct associated with them, as illustrated by Kopp (1987): "Thus it is permissible to shout in a playground but not a classroom, to run across a meadow but not a street, and to respect another's possessions whether the person is present or absent" (p. 34). The emergence of flexible adaptation is the hallmark of true regulatory competence.

Because early childhood is a time of critical development in self-regulation, it also is a time of special vulnerability for children at risk (Olson, Sameroff, Lunkenheimer, & Kerr, in press; Shonkoff & Phillips, 2000). Stable individual differences in ability to regulate affect and impulses can be detected in the first 3 years of life (National Institute of Child Health and Development Early Child Care Research Network, 2004) and predict adjustment outcomes across major developmental transitions—in some studies to late adolescence and early adulthood (Moffitt, Caspi, Harrington, & Milne, 2002; Olson, Schilling, & Bates, 1999).

We have highlighted the first 6 years of life, but concepts of self-regulation have important implications for understanding behavioral adjustment across the life span. For example, Dahl and Hariri (2004; see also Dahl & Conway, in press) have emphasized that adolescence is another time of special risk for the exacerbation of self-regulation deficits that underpin behavioral maladjustment. Like the preschool period, this period is defined by high levels of cortical growth (Sowell, Trauner, Gamst, & Jernigan, 2002). Unlike the preschool period, however, most parents tend to retreat from monitoring and structuring their teenager's behavioral routines. In vulnerable adolescents, this can easily set the scene for an escalation of risky behaviors that harm self and others (Dahl & Conway, in press).

In sum, failures in self-regulation are key mechanisms in the epigenesis of child behavior problems. We have described three definitional components of self-regulation that offer useful tools for understanding the development of both competence and psychopathology: integration across multiple levels (biological, behavioral, social), dynamic microprocesses that define an individual's responses to real-time challenges and perturbations, and cumulative and changing contexts over developmental time. The role of the interpersonal or social context is paramount in all of these self-regulatory components, especially with respect to early childhood, when crucial individual differences in self-regulation first emerge. In what follows, we show how these concepts can be used to enhance psychologists' understanding of a fundamental research issue: the nature of coregulatory exchanges between parents and children that exacerbate or ameliorate early disruptive problem behavior.

PARENT–CHILD COREGULATION OF AFFECT AS A TRANSACTIONAL PROCESS

The transactional model is an approach that illustrates the relations between developmental processes (Sameroff & MacKenzie, 2003) such as the relation between individual and dyadic development. A transaction between components of a system (e.g., parent and child) begins with a qualitative or quantitative change, but subsequently this new dynamic between elements may evolve into a stable pattern of behavioral organization (Sameroff & Chandler, 1975a). Thus, the relationship between parent and child is an active, self-organizing system in which stable patterns emerge over time from ongoing parent–child transactions. Furthermore, individual styles and circular dyadic patterns between parent and child are products of past recurring dyadic interactions, which potentially constrain qualities of future interactions (Cicchetti & Toth, 1997; Dumas & LaFreniere, 1995). For example, recurring aversive interactions between parent and child can create a rigid, maladaptive behavioral pattern that becomes increasingly difficult to alter (Hollenstein, Granic, Stoolmiller, & Snyder, 2004). Accordingly, early parent–child interaction has been viewed as a crucible for the development of long-standing patterns of behavioral adjustment.

Links between parent–child interaction and child behavior problems, in turn, are thought to reflect failures in the development of self-regulation. Researchers agree that early parent–child interaction is a primary context for the development of self-regulation and dysregulation (Sameroff & Emde, 1989; Sroufe, 1996). Generally, negative parent–child interaction hampers children's ability to regulate emotion (Eisenberg, Fabes, & Murphy, 1996; Gottman & Katz, 2002) and leads to increased behavioral problems (Olson, Bates, Sandy, & Schilling, 2002; Patterson, Reid, & Dishion, 1992). Conversely, positive parenting behaviors such as proactive structuring, responsiveness, and limit setting protect children from developing regulatory difficulties (Denham et al., 2000; Gottman, Katz, & Hooven, 1996) and have been linked to high levels of behavioral adjustment (Eisenberg et al., 2005). Furthermore, research with school-age children has shown that children's emotion regulation mediates the effects of positive and negative parenting on children's later behavior problems (Chang, Schwartz, Dodge, & McBride-Chang, 2003; Eisenberg et al., 2003).

Frequently, individual assessments of child and parent characteristics have been used to examine the dynamic relationship between them. A test of mediation reflects both parents' and children's contributions, but it still implies linear and separate causal arrows and is variable-centered, treating regulation as an individual trait rather than a dynamic process. During dyadic interactions, child and parent engage in a fluid, circular interchange whereby the child prompts the parent to respond and the parent's response alters the child's behavior (Maccoby, 1992). Additionally, parent–child emotional exchanges reflect structural aspects of relationships and not just the content therein. For example, a particular pattern of parent–child interaction behavior (e.g., restricted affective flexibility) may support or hinder child development over time to the same extent as the content of the episode (Harrist & Waugh, 2002; Hollenstein et al., 2004). To understand the role of self-regulation in the etiology of child behavior problems, we must directly study the temporal unfolding and patterning of parent–child emotion exchanges (Granic, 2000).

In what follows, we explore how the study of affect coregulation, assessed in both real time (microscopic lens) and developmental time (macroscopic lens), offers promise as one way of examining transactional processes in development.

Affect Regulation as an Interpersonal Process

The development of self-regulation strongly reflects mutual dyadic exchanges between the child and his or her caregivers (Fogel, 1993; Harrist & Waugh, 2002; Mikulincer & Shaver, 2004). The coregulation of affect begins in infancy as caregiver and infant match each other's affective states in close temporal proximity, thus jointly regulating the infant's positive arousal during face-to-face interactions (Feldman, 2003). Dyadic coregulatory interactions continue into the toddler and preschool years, taking different forms concomitant with the child's developmental stage (Harrist & Waugh, 2002). For example, parents and their preschoolers often must coregulate negative affect in the context of frustration, such as when a child will not comply with a parental request. Although such interactions are common in the preschool years, the study of coregulation as a dynamic process between parent and child has received little empirical attention outside of the mother–infant relationship (Cole, Teti, & Zahn-Waxler, 2003).

Coregulatory processes have been described using a variety of constructs such as parent–child dyadic coregulation (Fogel, 1993), mutual regulation (Cole et al., 2003; Tronick, 1989), dyadic or positive synchrony (Criss, Shaw, & Ingoldsby, 2003; Harrist, Pettit, Dodge, & Bates, 1994), and dyadic mutuality (Deater-Deckard, Atzaba-Poria, & Pike, 2004). Specific aspects of coregulation that have been studied include contingency (Cole et al., 2003; Dumas, LaFreniere, & Serketich, 1995), responsiveness (Deater-Deckard et al., 2004; Kochanksa & Aksan, 2004), behavioral/affective flexibility (Granic, Hollenstein, Dishion, & Patterson, 2003; Hollenstein et al., 2004), and temporal coordination of affect (Feldman, Greenbaum, & Yirmiya, 1999; Tronick & Cohn, 1989). Likewise, the level of analysis has differed, with both macro- and microlevel coding of coregulation showing associations with child outcomes. Researchers have touted the advantages of using macro- and microlevel analysis conjointly (e.g., Maccoby & Martin, 1983), and this interplay is especially relevant in the discussion of how microscopic interactions in real time may develop into stable patterns of dyadic interaction across development (Dumas, Lemay, & Dauwalder, 2001). Consequently, research on coregulation rests at the intersection of overlapping definitions, theories, and methodologies.

Although theoretical conceptualizations of coregulation differ, *dyadic synchrony* is generally defined as "an observable pattern of dyadic interaction that is mutually regulated, reciprocal, and harmonious" (Harrist & Waugh, 2002, p. 557). Thus, coregulation characterizes a style or pattern of interaction, whereas synchrony reflects the adaptive aspects of this process. Some theorists include the matching of affect in their definition of synchrony (Stern, 1985; Tronick & Gianino, 1986); some include the coordination of positive affect specifically (Censullo, Bowler, Lester, & Brazelton, 1987); and others believe that it is important to represent affect and synchrony separately to learn more about individual differences among dyads (Kirsh, Crnic, & Greenberg, 1995). Synchronous interactions engender positive emotions, whereas asynchronous interactions obstruct them. Thus, regardless of the specific definition used, dyadic synchrony and the coregulation of emotion are intimately related constructs, if not synonymous in some cases.

Developmental Changes in Parent–Child Coregulation

As described previously, parent–child coregulation begins in infancy with the parent's aiding and supplementing the infant's goal-directed strivings through a joint affective communication system (Fogel, 1993; Tronick, 1989). Theorists generally agree that there is a developmental path from coordinated dyadic interactions at 3 months to the child's self-regulation at 3 years (Kopp, 1982). Feldman et al. (1999) found evidence for this theoretical pathway. They defined coregulation as the temporal coordination of changes (on the basis of stable time lags) in affectively based mother–infant play. Mother–infant coregulation at 3 and 9 months and the child's regulatory ability at 2 years were significantly intercorrelated. Moreover, synchronous coregulation increased with age. Tronick and Cohn (1989) found that mother–infant pairs spent much of their play time in miscoordinated affective states, but that these were typically corrected in the next time interval. Mother–infant matching of affective states improved across 3, 6, and 9 months, although synchrony (changing conjointly during interaction) did not. Kochanska and Aksan (2004) used time series analyses to derive behavioral indexes of mutual mother–infant responsiveness at 7 and 15 months. Behavioral measures of mutual responsiveness showed no relation over time, but global ratings of parent and infant responsiveness at 7 months each predicted the partner's responsiveness at 15 months (Kochanska & Aksan, 2004).

In sum, studies indicate that coregulation between parent and infant generally improves over time. Thus, one question that arises from infant studies is whether one can expect stability in dyadic coregulation across the toddler and preschool years. Stability of coregulation could change in early childhood when developmental capabilities and demands are different and the coordination of the child's regulatory systems is more refined. Identifying sources of stability and instability in dyadic coregulation may inform psychologists' understanding of pathways to diverse child adjustment outcomes in later years.

Coregulation is thought to serve different purposes as the child becomes a more equal partner in parent–child exchanges, for example, by sharing in turn taking, initiatives, and overall engagement (Harrist & Waugh, 2002;

Kochanska & Aksan, 2004). Synchronous interactions between toddlers and their parents allow the child to prepare for interactions with peers through improved communication skills and practice with autonomy (Harrist & Waugh, 2002; Kirsh et al., 1995). Indeed, children who experience more synchronous and mutually responsive parent–child interactions have been found to show higher levels of compliance and social competence (Harrist et al., 1994; Lindsey, Mize, & Pettit, 1997), moral conscience (Kochanska & Murray, 2000), and cognitive development (Kirsh et al., 1995) than others.

Quality of parent–child coregulation becomes particularly important in the preschool years when chronic conduct problems develop (Campbell, 2002). In preschool, conflictual and mutually negative parent–child interactions become more common and are thought to be potential bellwethers of long-term behavior problems (Harrist & Waugh, 2002). For example, coercive interaction patterns between parent and child begin to stabilize during the preschool period (Patterson, Capaldi, & Bank, 1991), distinguish preschoolers with externalizing symptoms from typical children (Dumas et al., 1995), and predict the development of conduct disorder (Patterson et al., 1991). For these reasons, the preschool period can be considered a crucible for the development of persistent patterns of child regulatory difficulties (Cole et al., 2003). Thus, it is particularly important to understand the developmental progression of individual differences in parent–child coregulation and their relations to child problem behavior across the preschool years. We now describe work that has addressed this question and highlight the various methods that have been used to assess parent-child coregulation.

Multilevel Assessments of Coregulation in Early Childhood

In some studies, quality of parent–child coregulation has been assessed using global ratings (Deater-Deckard et al., 2004; Harrist et al., 1994; Mize & Pettit, 1997). Global ratings of mother–child dyadic synchrony have been linked with low levels of child aggression in kindergarten (Harrist et al., 1994) and preschool (Mize & Pettit, 1997) settings. Research in later childhood has mirrored these findings. For example, global ratings of parent–child reciprocity (e.g., joint attention, co-occurring positive affect, and eye contact), responsiveness (e.g., frequency and latency of child responses to parent and parent responses to child), and cooperation (e.g., explicit agreement and discussion about how to proceed with the task) were found to be negatively associated with children's externalizing problems at ages 7 to 9 years (Deater-Deckard et al., 2004).

Research using sequential time series assessments also has demonstrated promising findings with regard to the importance of coregulatory processes in development. For example, Dumas et al. (1995) analyzed mother–child interactions and their relationship to children's socially competent, anxious, and aggressive group classifications. Specifically, they assessed the frequency of parent–child exchanges that began with a control attempt by either partner, and coded initiations and responses for positive and negative affect, parental control, and child compliance. Competent children and their mothers showed more positive reciprocal chains, with firm limit setting in the face of coercive child behaviors. Aggressive children made more coercive control attempts than others, and their mothers responded by failing to set limits and by showing indiscriminant affective responses (i.e., they were equally likely to show positive or negative affect). Although cross-sectional, this research revealed that mutual, contingent positive affect in parent–child interaction was associated with greater child competence. Moreover, firm parental responses to child coercion that occurred in the context of reciprocal positive affect differentiated competent children from others. However, mutual negativity was not always characteristic of mother–child dyads with aggressive children as would be expected on the basis of prior theoretical accounts (Patterson et al., 1991).

Cole et al. (2003) also used microlevel analyses to examine affective coregulation in a community sample of 5-year-old children and their mothers. During a laboratory waiting task, participants' facial and vocal expressions were coded second by second for the presence of positive and angry affect; coregulation was then defined as the frequency of self-initiated and contingent positive and angry emotions for each partner. Mothers' contingent angry emotional responses predicted stability in children's externalizing problems in school settings between 5 and 7 years of age. Conversely, children whose externalizing behavior improved over time had mothers who made more positive contingent responses. These children also made more neutral contingent responses than did children with persistent problems.

A different approach to defining quality of parent–child coregulation was used by Hollenstein et al. (2004), whose work reflects the influence of dynamic systems principles. Dynamic systems approaches to coregulation reflect the structural dynamics of parent–child interaction, for example, the flexibility of the affective repertoire, the predictability of behavioral patterns, or the tendency to return to a specific dyadic state (see Lunkenheimer & Dishion, 2008). Hollenstein et al. questioned whether structural qualities of parent–child interaction could serve as markers of children's behavioral adjustment over the course of the kindergarten year. Participants were three consecutive cohorts of kindergarten-age children from one elementary school in a low-income neighborhood. State space grid analysis (Lewis, Lamey, & Douglas, 1999) was used to derive the range of affective behaviors expressed, the number of transitions from one type of affect to another, and the average time spent in each type of affect in a parent–child assessment at the beginning of the year. Lewis et al. (1999) hypothesized that poorly coregulated dyads would demonstrate a diminished behavioral repertoire, a tendency to avoid change, and a tendency toward perseveration of certain affective states. Indeed, early dyadic rigidity predicted the status of children who showed consistently high and increasing levels of problem behavior across the kindergarten year. Although research has demonstrated that coercive parent–child relations are in place as early as 24 months (Shaw et al., 1998), this study showed that structural patterns of parent–child interaction continued to mark child adjustment problems after the transition to school. Moreover, the impact of patterns of coregulation remained even after controlling for the content of the interaction, highlighting the importance of including structural analysis in future research.

Michigan Longitudinal Study

Most previous studies have focused exclusively on mother–child dyads. Through an expanded study of parent–child coregulation to include fathers as well, our recent work also has explored patterns of affective coregulation as predictors of children's later behavioral adjustment outcomes (Lunkenheimer, Olson, Kaciroti, & Sameroff, 2007). The Michigan Longitudinal Study is a prospective longitudinal investigation of the development of children's externalizing problems across the transition period between early preschool and school entry (Olson et al., 2005). When children were 3 years of age, mother–child and father–child interactions were videotaped on separate days in the family's home. Parents and children were asked to complete a series of challenging structured tasks (e.g., puzzles that were too difficult for the child to solve independently). On the basis of continuous records of moment-to-moment affective exchanges between parents and children, sequential analysis was used to derive constructs of coregulation in mother– and father–child dyads. Coregulation was defined as the correlation between partner behaviors at one time lag (e.g., parent positive affect at t and child positive affect at $t+1$); these correlations were then averaged across the length of the interaction. Both directions of influence (i.e., time-lagged relationships from child to parent and from parent to child) were assessed. Longitudinal structural equation models revealed that preschoolers who typically responded to parents' positive affect with an increase in their own positive affect showed fewer externalizing problems across the transition to kindergarten as rated by mothers, fathers, and teachers (see Figure 4.1). Predictive pathways between early affective coregulation and children's later behavioral adjustment were evident even when prior influences of family socioeconomic status, child gender, and child temperament were accounted for in the model. Notably, children's average levels of positive affect alone did not predict later outcomes. Thus, patterns of affective coregulation between parents and their young children were linked with individual differences in children's externalizing behavior problems across a critical developmental transition.

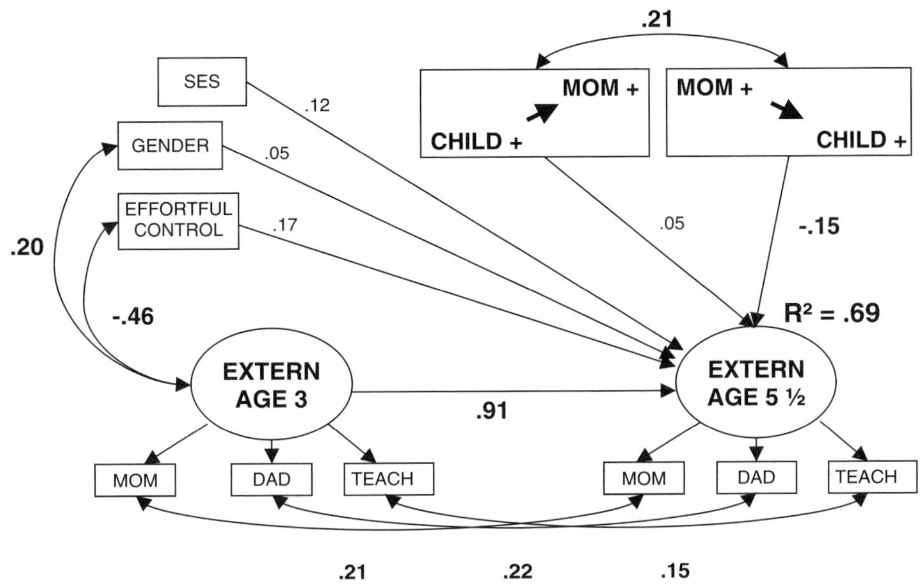

Figure 4.1. Structural equation model of the effects of mother–child coregulation of positive affect and control variables at age 3 years on children's externalizing behavior problems at age 5 1/2 years (Lunkenheimer et al. 2007). SES = socioeconomic status; EXTERN = externalizing behaviors, TEACH = teacher, CHILD+ = child positive affect; MOM+ = mother positive affect.

SUMMARY AND CONCLUSIONS

We began with a powerful theoretical framework, now over 30 years old, which has become deeply embedded into the ways psychologists think about development, mental health, and intervention. Ironically, however, translating the central tenets of transactional theory into empirical research paradigms has continued to be extremely challenging (e.g., Sameroff & MacKenzie, 2003). This is especially true in relation to the unique defining feature of the transactional model: the presence of transformational child–environment interactions across time. With their inherent dynamic and multisystemic properties, concepts of self-regulation provide a family of tools for examining the nature of transformational processes in development.

In this chapter we have explored how individually focused constructs of affective and behavioral self-regulation can be expanded into constructs that capture complex coregulatory processes between parents and their young children. Recent studies in our own laboratory and in others' suggest that qualities of parent–child coregulation are important markers of children's later behavioral development, particularly the ability to self-regulate angry affect and aggressive impulses. Establishing the link between regulatory processes in real time and children's later behavioral adjustment is an important step in the direction of representing dynamic, transactional processes across development (Lunkenheimer & Dishion, in press). A working hypothesis, yet to be tested in further longitudinal work, is that constructs of coregulation offer one way of representing mutual influence processes that may become internalized into traitlike behavioral deficits and competencies.

Possibilities for further research are exciting. A full model of coregulatory processes would encompass not only moment-to-moment variation in affective synchrony but also the actors' cognitive representations of their relationships. For example, the same noncompliant child behavior may be perceived as relatively benign by one parent and as a sign of rejection by another (see chap. 6, this volume). Moreover, we would expect shifts in quality of coping with interpersonal microstressors to reflect not only significant qualities of dyadic relationship systems but also the weight and chronicity of the family stress load.

Finally, our journey began with the challenge of translating key tenets of transactional theory into measurement constructs. We conclude that a necessary foundation for the translation process is the presence of observational data containing moment-to-moment variations in actors' responses to one another as they unfold across real time and developmental time. Despite methodological difficulties inherent in following this path, a true representation of transactional processes in development hinges on paying close attention to interpersonal exchanges as they ebb and flow across time and, ultimately, to the particulars of individual lives.

REFERENCES

Bell, R. Q. (1971). Stimulus control of parent or caretaker behavior by offspring. *Developmental Psychology, 4,* 63–72.

Calkins, S. D. (in press). Regulatory competence and early disruptive behavior: The role of physiological regulation. In S. L. Olson & A. J. Sameroff (Eds.), *Regulatory processes in the development of child behavior problems*. New York: Cambridge University Press.

Campbell, S. B. (2002). *Behavior problems in preschool children: Clinical and developmental issues*. New York: Guilford Press.

Campbell, S. B., Shaw, D. S., & Gilliom, M. (2000). Early externalizing behavior problems: Toddlers and preschoolers at risk for later maladjustment. *Development and Psychopathology, 12,* 467–488.

Censullo, M., Bowler, R., Lester, B., & Brazelton, T. B. (1987). An instrument for the measurement of infant–adult synchrony. *Nursing Research, 36,* 244–248.

Chang, L., Schwartz, D., Dodge, K. A., & McBride-Chang, C. (2003). Harsh parenting in relation to child emotion regulation and aggression. *Journal of Family Psychology, 17,* 598–606.

Charles, S. T., & Carstensen, L. L. (2007). Emotion regulation and aging. In J. Gross (Ed.), *Handbook of emotion regulation* (pp. 307–327). New York: Guilford Press.

Cicchetti, D., & Toth, S. L. (1997). Transactional ecological systems in developmental psychopathology. In S. S. Luthar & J. A. Burack (Eds.), *Developmental psychopathology: Perspectives on adjustment, risk, and disorder* (pp. 317–349). New York: Cambridge University Press.

Cole, P. M., Martin, S. E., & Dennis, T. A. (2004). Emotion regulation as a scientific construct: Methodological challenges and directions for child development research. *Child Development, 75*, 317–333.

Cole, P. M., Teti, L. O., & Zahn-Waxler, C. (2003). Mutual emotion regulation and the stability of conduct problems between preschool and school age. *Development and Psychopathology, 15*, 1–18.

Criss, M. M., Shaw, D. S., & Ingoldsby, E. M. (2003). Mother–son positive synchrony in middle childhood: Relation to antisocial behavior. *Social Development, 12*, 379–400.

Dahl, R. E. & Conway, A. M. (in press). Self-regulation and the development of behavioral and emotional problems: Toward an integrative conceptual and translational research agenda. In S. L. Olson & A. J. Sameroff (Eds.), *Regulatory processes in the development of child behavior problems*. New York: Cambridge University Press.

Dahl, R. E. & Hariri, A. R. (2005). Lessons from G. Stanley Hall: Connecting new research in the biological sciences to the study of adolescent development. *Journal of Research on Adolescence, 15*, 367–382.

Deater-Deckard, K., Atzaba-Poria, N., & Pike, A. (2004). Mother– and father–child mutuality in Anglo and Indian British families: A link with lower externalizing problems. *Journal of Abnormal Child Psychology, 32*, 609–620.

Denham, S. A., Workman, E., Cole, P. M., Weissbrod, C., Kendziora, K. T., & Zahn-Waxler, C. (2000). Prediction of externalizing behavior problems from early to middle childhood: The role of parental socialization and emotion expression. *Development and Psychopathology, 12*, 23–45.

Diamond, L. M., & Aspinwall, L. G. (2003). Emotion regulation across the life span: An integrative perspective emphasizing self-regulation, positive affect, and dyadic processes. *Motivation and Emotion, 27*, 125–156.

Dumas, J. E., & LaFreniere, P. J. (1995). Relationships as context: Supportive and coercive interactions in competent, aggressive, and anxious mother–child dyads. In J. McCord (Ed.), *Coercion and punishment in long-term perspectives* (pp. 9–33). New York: Cambridge University Press.

Dumas, J. E., LaFreniere, P. J., & Serketich, W. J. (1995). "Balance of power": A transactional analysis of control in mother–child dyads involving socially competent, aggressive, and anxious children. *Journal of Abnormal Child Psychology, 104*, 104–113.

Dumas, J. E., Lemay, P., & Dauwalder, J. (2001). Dynamic analyses of mother–child interactions in functional and dysfunctional dyads: A synergetic approach. *Journal of Abnormal Child Psychology, 29*, 317–329.

Eisenberg, N., Fabes, R. A., & Murphy, B. C. (1996). Parents' reactions to children's negative emotions: Relations to children's social competence and comforting behavior. *Child Development, 67*, 2227–2247.

Eisenberg, N., Valiente, C., Morris, A. S., Fabes, R. A., Cumberland, A., Reiser, M., et al. (2003). Longitudinal relations among parental emotional expressivity, children's regulation, and quality of socioemotional functioning. *Developmental Psychology, 39*, 3–19.

Eisenberg, N., Zhou, Q., Spinrad, T. L., Valiente, C., Fabes, R. A., & Liew, J. (2005). Relations among positive parenting, children's effortful control, and externalizing problems: A three-wave longitudinal study. *Child Development, 76*, 1055–1071.

Feldman, R. (2003). Infant–mother and infant–father synchrony: The coregulation of positive arousal. *Infant Mental Health Journal, 24*, 1–23.

Feldman, R., Greenbaum, C. W., & Yirmiya, N. (1999). Mother–infant affect synchrony as an antecedent of the emergence of self-control. *Developmental Psychology, 35*, 223–231.

Fogel, A. (1993). Two principles of communication: Co-regulation and framing. In J. Nadel & L. Camaioni (Eds.), *New perspectives in early communicative development* (pp. 9–22). London: Routledge.

Forbes, E. E., Fox, N. A., Cohn, J. F., Galles, S. F., & Kovacs, M. (2006). Children's affect regulation during a disappointment: Psychophysiological responses and relations to parent history of depression. *Biological Psychiatry, 71*, 264–277.

Fox, N. A, & Calkins, S. D. (2003). The development of self-control of emotion: Intrinsic and extrinsic influences. *Motivation and Emotion, 27* (1), 7–26.

Gottman, J. M., & Katz, L. F. (2002). Children's emotional reactions to stressful parent–child interactions: The link between emotion regulation and vagal tone. *Marriage & Family Review, 34*, 265–283.

Gottman, J. M., Katz, L. F., & Hooven, C. (1996). Parental meta-emotion philosophy and the emotional life of families: Theoretical models and preliminary data. *Journal of Family Psychology, 10*, 243–268.

Granic, I. (2000). The self-organization of parent–child relations: Beyond bidirectional models. In M. D. Lewis & I. Granic (Eds.), *Emotion, development, and self-organization: Dynamic systems approaches to emotional development* (pp. 267–297). New York: Cambridge University Press.

Granic, I., Hollenstein, T., Dishion, T. J., & Patterson, G. R. (2003). Longitudinal analysis of flexibility and reorganization in early adolescence: A dynamic systems study of family interactions. *Developmental Psychology, 39*, 606–617.

Harrist, A. W., Pettit, G. S., Dodge, K. E., & Bates, J. E. (1994). Dyadic synchrony in mother–child interaction: Relation with children's subsequent kindergarten adjustment. *Family Relations, 43*, 417–424.

Harrist, A. W., & Waugh, R. M. (2002). Dyadic synchrony: Its structure and function in children's development. *Developmental Review, 22*, 555–592.

Hollenstein, T., Granic, I., Stoolmiller, M., & Snyder, J. (2004). Rigidity in parent–child interactions and the development of externalizing and internalizing behavior in early childhood. *Journal of Abnormal Child Psychology, 32*, 595–607.

Keenan, K. (2000). Emotion dysregulation as a risk factor for child psychopathology. *Clinical Psychology: Science and Practice, 7*, 418–434.

Keenan, K., Jacob, S., Grace, G., & Gunthorpe, D. (in press). Context matters: Exploring definitions of a poorly modulated stress responses. In S. L. Olson & A. J. Sameroff (Eds.), *Regulatory processes in the development of child behavior problems*. New York: Cambridge University Press.

Kirsh, S. J., Crnic, K. A., & Greenberg, M. T. (1995). Relations between parent–child affect and synchrony and cognitive outcome at 5 years of age. *Personal Relationships, 2*, 187–198.

Kochanska, G., & Aksan, N. (2004). Development of mutual responsiveness between parents and their young children. *Child Development, 75*, 1657–1676.

Kochanska, G., & Murray, K. T. (2000). Mother–child mutually responsive orientation and conscience development: From toddler to early school age. *Child Development, 71*, 417–431.

Kopp, C. B. (1982). Antecedents of self-regulation: A developmental perspective. *Developmental Psychology, 18*, 199–214.

Kopp, C. B. (1987). Developmental risk: Historical reflections. In J. D. Osofsky (Ed.), *Handbook of infant development* (2nd ed., pp. 881–912). Oxford, England: Wiley.

Kopp, C. B. (1989). Regulation of distress and negative emotion: A developmental view. *Developmental Psychology, 25*, 343–354.

Lewis, M. D., Lamey, A. V., & Douglas, L. (1999). A new dynamic systems method for the analysis of early socioemotional development. *Developmental Science, 2*, 457–475.

Lindsey, E. W., Mize, J., & Pettit, G. S. (1997). Mutuality in parent–child play: Consequences for children's peer competence. *Journal of Social and Personal Relationships, 14*, 523–538.

Lopez, N. L. (2006). *Individual differences in the activation and regulation of the limbic–hypothalamic–pituitary–adrenal axis after stress: Implications for emotion regulation*. Ann Arbor, MI: Proquest Information and Learning.

Lopez, N. L., Vazquez, M., & Olson, S. L. (2004). An integrative approach to the neurophysiological substrates of social withdrawal and aggression. *Development and Psychopathology, 16*, 69–93.

Lopez-Duran, N. L., Olson, S. L., Felt, B., & Vazquez, M. (in press). An integrative approach to the neurophysiology of emotion regulation: The case of social withdrawal. In S. L. Olson & A. J. Sameroff (Eds.), *Regulatory processes in the development of child behavior problems*. New York: Cambridge University Press.

Lunkenheimer, E. S., & Dishion, T. J. (2008). Developmental psychopathology: Maladaptive and adaptive attractors in children's close relationships. In S. Guastello, M. Koopmans, & D. Pincus (Eds.), *Chaos and complexity in psychology: The theory of nonlinear dynamical systems* (pp. 282–306). New York: Cambridge University Press.

Lunkenheimer, E. S., Olson, S. L., Kaciroti, N., & Sameroff, A. J. (2007, March). *Parent–child coregulation of affect in early childhood and children's behavior problems across the transition to school*. Paper presented at the Biennial Meeting of the Society for Research in Child Development, Boston.

Maccoby, E. E. (1992). The role of parents in the socialization of children: An historical overview. *Developmental Psychology*, *28*, 1006–1017.

Maccoby, E. E., & Martin, J. A. (1983). Socialization in the context of the family: Parent–child interaction. In P. H. Mussen (Series Ed.) & E. M. Hetherington (Vol. Ed.), *Handbook of child psychology: Vol. 4. Socialization, personality, and social development* (pp. 1–102). New York: Wiley.

Mikulincer, M., & Shaver, P. R. (2004). Security-based self-representations in adulthood: Contents and processes. In W. S. Rholes & J. A. Simpson (Eds.), *Adult attachment: Theory, research, and clinical implications* (pp. 159–195). New York: Guilford Press.

Miller, A. L., & Olson, S. L. (2000). Emotional expressiveness during peer conflicts: A predictor of social maladjustment among high-risk preschoolers. *Journal of Abnormal Child Psychology*, *28*(4), 339–352.

Mize, J., & Pettit, G. S. (1997). Mothers' social coaching, mother–child relationship style, and children's peer competence: Is the medium the message? *Child Development*, *68*, 312–332.

Moffitt, T. E., Caspi, A., Harrington, H., & Milne, B. J. (2002). Males on the life-course-persistent and adolescence-limited antisocial pathways: Follow-up at age 26 years. *Development and Psychopathology*, *14*, 179–207.

National Institute of Child Health and Development Early Child Care Research Network. (2004). Affect dysregulation in the mother–child relationship in the toddler years: Antecedents and consequences. *Development and Psychopathology*, *16*, 43–68.

Olson, S. L. (1992). Development of conduct problems and peer rejection in preschool children: A social systems analysis. *Journal of Abnormal Child Psychology*, *20*(3), 327–350.

Olson, S. L., Bates, J. E., Sandy, J. M., & Schilling, E. M. (2002). Early developmental precursors of impulsive and inattentive behavior: From infancy to middle childhood. *Journal of Child Psychology and Psychiatry*, *43*, 435–447.

Olson, S. L., Sameroff, A. J., Kerr, D. C. R., Lopez, N. L., & Wellman, H. M. (2005). Developmental foundations of externalizing problems in young children: The role of effortful control. *Development and Psychopathology*, *17*, 25–45.

Olson, S. L., Sameroff, A. J., Lunkenheimer, E. S., & Kerr, D. C. R. (in press). Self-regulatory processes in early behavioral adjustment: The preschool to school transition. In S. L. Olson & A. J. Sameroff (Eds.), *Regulatory processes in the development of behavior problems: Biological, behavioral, and social-ecological interactions*. New York: Cambridge University Press.

Olson, S. L., Schilling, E. M., & Bates, J. E. (1999). Measurement of impulsivity: Construct coherence, longitudinal stability, and relationships with externalizing problems in middle childhood and adolescence. *Journal of Abnormal Child Psychology*, *27*, 151–165.

Patterson, G. R., & Bank, L. (1989). Some amplifying mechanisms for pathologic processes in families. In M. R. Gunnar & E. Thelen (Eds.), *Systems and development* (pp. 167–209). Hillsdale, NJ: Erlbaum.

Patterson, G. R., Capaldi, D., & Bank, L. (1991). An early starter model for predicting delinquency. In D. J. Pepler & K. H. Rubin (Eds.), *The development and treatment of childhood aggression* (pp. 139–168). Hillsdale, NJ: Erlbaum.

Patterson, G. R., Reid, J., & Dishion, T. J. (1992). *Antisocial boys*. Eugene, OR: Castalia.

Posner, M. I., & Rothbart, M. K. (2000). Developing mechanisms of self-regulation. *Development and Psychopathology, 12*, 427–441.

Rothbart, M. K., Sheese, B. E., & Posner, M. I. (2007). Executive attention and effortful control: Linking temperament, brain networks, and genes. *Child Development Perspectives, 1*, 1–63.

Rueda, M. R., Posner, M. I., & Rothbart, M. K. (2005). The development of executive attention: Contributions to the emergence of self-regulation. *Development and Neuropsychology, 28*, 573–594.

Sameroff, A. J., & Chandler, M. (1975a). Early influences on development: Fact or fancy? *Merrill-Palmer Quarterly, 21*, 267–294.

Sameroff, A. J., & Chandler, M. J. (1975b). Reproductive risk and the continuum of caretaking casualty. In F. Horowitz (Ed.), *Review of child development research* (pp. 187–244). Chicago: University of Chicago Press.

Sameroff, A. J., & Emde, R. N. (1989). *Relationship disturbances in early childhood*. New York: Basic Books.

Sameroff, A. J., & MacKenzie, M. J. (2003). Research strategies for capturing transactional models of development: The limits of the possible. *Development and Psychopathology, 15*, 613–640.

Shaw, D. S., Winslow, E. B., Owens, E. B., Vondra, J. J., Cohn, J. E., & Bell, R. Q. (1998). The development of early externalizing problems among children from low-income families: A transformational perspective. *Journal of Abnormal Child Psychology, 26*, 95–107.

Shonkoff, J. P., & Phillips, D. A. (2000). *From neurons to neighborhoods: The science of early childhood development*. Washington, DC: National Academies Press.

Sowell, E. R., Trauner, D. A., Gamst, A., & Jernigan, T. L. (2002). Development of cortical and subcortical brain structures in childhood and adolescence: A structural MRI study. *Developmental Medicine & Child Neurology, 44*, 4–16.

Sroufe, L. A. (1996). *Emotional development: The organization of emotional life in the early years*. Cambridge, England: Cambridge University Press.

Stern, D. (1985). *The interpersonal world of the infant*. New York: Basic Books.

Thompson, R. A. (1994). Emotion regulation: A theme in search of definition. *Monographs of the Society for Research in Child Development, 59*(2–3 Serial No. 240).

Tremblay, R. E. (2000). The development of aggressive behavior during childhood: What have we learned in the past century? *International Journal of Behavioral Development, 24*, 129–141.

Tronick, E. Z. (1989). Emotions and emotional communication in infants. *American Psychologist, 44*, 112–119.

Tronick, E. Z., & Cohn, J. F. (1989). Infant–mother face-to-face interaction: Age and gender differences in coordination and the occurrence of miscoordination. *Child Development, 60*, 85–92.

Tronick, E. Z., & Gianino, A. (1986). Interactive mismatch and repair: Challenges to the coping infant. *Zero to Three*, 6, 1–6.

Volling, B. V., Kolak, A. M., & Blandon, A. Y. (in press). Family subsystems and children's self-regulation. In S. L. Olson & A. J. Sameroff (Eds.), *Regulatory processes in the development of child behavior problems*. New York: Cambridge University Press.

5

DEVELOPMENTAL TRANSACTIONS BETWEEN BOYS' CONDUCT PROBLEMS AND MOTHERS' DEPRESSIVE SYMPTOMS

DANIEL S. SHAW, HEATHER E. GROSS, AND KRISTIN L. MOILANEN

Despite the fact that child effects on parenting practices and subsequent child behavior have been postulated since Bell's (1968) seminal theoretical article, and at a broader level by Sameroff and Chandler's (1975) transactional perspective, relatively few studies have explored transactional processes between maternal depression and child adjustment over time. The goal of this chapter is to examine bidirectional processes in the relationship between maternal depression and child conduct problems. This work was inspired by the plethora of research examining unidirectional associations between maternal depression and multiple types of child outcomes and the relative dearth of research examining the possibility that such associations might be bidirectional. The current chapter's focus is on testing Sameroff's original transactional model, but instead of focusing on parenting behavior per se, we chose to examine a more distal parental factor that has been consistently linked to both caregiving practices and child outcomes, maternal depression (Belsky, 1984; Conger, Patterson, & Ge, 1995).

The research reported in this article was supported by grants to Daniel S. Shaw from the National Institute of Mental Health (MH 46925, MH 50907, and MH 01666). We are grateful to the staff of the Pitt Mother & Child Project for their years of service and to our study families for making the research possible.

Parental psychopathology has been found to be a consistent and robust correlate of children's maladjustment (Goodman & Brumley, 1990). Because of the prevalence of depression, especially in women, maternal depression has been the focus of numerous research studies on parental psychopathology and its association with child psychopathology. Findings in the extant literature provide substantial evidence for an association between maternal depression and negative child outcomes, including internalizing and conduct problems (for reviews of this literature, see Beardslee, Versage, & Gladstone, 1998; Cummings & Davies, 1994). In addition to research on the relationship between maternal depression and different forms of child psychopathology, associations have been found between child characteristics and parental behavior (Bell & Harper, 1977; Elgar, McGrath, Waschbusch, Stewart, & Curtis, 2004). Rather than consider parent effects on children and child effects on parents to be separate processes, reciprocal models of socialization regard parenting behaviors and child characteristics as recurrent transactional exchanges over time in which both parties affect the other (Bell, 1968; Sameroff, 1995). Although there is an extensive body of research on reciprocal effects between child disruptive behavior and aspects of parenting (Bell & Harper, 1977; Johnston & Mash, 2001), substantially less attention has been paid to potential bidirectional effects between child disruptive behavior and parental mental health, such as depressive symptoms, over time. Because both maternal clinical depression and subclinical elevated levels of depressive symptoms have been found to be related to child maladjustment (Cummings, Keller, & Davies, 2005), the term *maternal depression* is used throughout this chapter to describe both criteria. Similarly, the term *conduct problems* is used to describe a range of heterotypically similar externalizing symptoms, ranging from oppositional and aggressive behavior in early and middle childhood to more covert antisocial activities beginning in late middle childhood.

There have been consistent findings linking maternal depression to disruptions in both socioemotional and instrumental functioning (Elgar, McGrath, Waschbusch, Stewart, & Curtis, 2004), when children of depressed mothers are studied across both narrowly defined developmental periods and broad age spans (Goodman & Gotlib, 1999). For example, during early childhood, maternal depression has been linked to fussiness and difficult child temperament (Whiffen & Gotlib, 1989), insecure attachment (Campbell et al., 2004), conduct problems (Marchand, Hock, & Widaman, 2002), and reduced mental and motor development (Murray, Fiori-Cowley, Hooper, & Cooper, 1996). Similarly, studies of school-age children and adolescents have documented associations between maternal depression and elevated rates of internalizing and externalizing problems, more serious forms of antisocial behavior (Hay, Pawlby, Angold, Harold, & Sharp, 2003), lower levels of self-esteem (Cummings et al., 2005), and academic problems (Sinclair & Murray, 1998).

Child Effects on Parents and Reciprocal Models

Child effects models emphasize the influence of children's attributes and behaviors on their parents. The literature on parenting is replete with theoretical and empirical evidence of child effects on parents. Belsky's (1984) landmark article on the determinants of parenting provides a foundation for reciprocal parenting models by positing that characteristics of both the parent and child contribute to adaptive and dysfunctional parenting. This idea is expanded in Patterson's (1982) coercive model of parenting in which a cycle of negative reinforcement is established when child noncompliance is reinforced by the parent. Accordingly, parents unwittingly reinforce a child's disruptive behavior by paying more attention to it than to a child's prosocial behavior (Eddy, Leve, & Fagot, 2001). These types of coercive parenting practices have been linked to long-term difficulties for children, particularly in rates of conduct problems (Campbell, Shaw, & Gilliom, 2000).

Just as child behaviors are thought to influence parenting, a number of studies have found evidence for child effects on other adult behaviors, including marital quality (Leve, Scaramella, & Fagot, 2001), alcohol consumption (Pelham et al., 1997), social life (Donenberg & Baker, 1993), parenting self-efficacy (Teti & Gelfand, 1991), and stress (Baker & Heller, 1996). Moreover, there is a growing body of research on child effects and maternal depression. Coyne's interpersonal model of depression provides a theoretical basis for bidirectional effects of depression in describing how depressed adults elicit negative reactions from others that intensify their unhappiness and negativity in a cycle of mutual distress (Coyne, Kahn, & Gotlib, 1987). Nelson, Hammen, Brennan, and Ullman (2003) speculated that maternal depression may create dysfunctions in the early parent–child relationship and elicit problems in the child, which would, in turn, maintain negative maternal attitudes.

Research findings that support a child effects model of maternal depression include higher rates of maternal depressive symptoms in samples of clinic-referred versus normal children (Fergusson, Lynskey, & Horwood, 1993) and in mothers whose children have high levels of behavioral or emotional problems (Civic & Holt, 2000). Infancy is an interesting period in which to examine potential child effects because of the young child's dependence on parents coupled with the child's inability to communicate his or her needs verbally. For example, Field et al. (1988) found that when infants of depressed mothers interacted with nondepressed adults in avoidant and unresponsive ways, nondepressed adults began to exhibit depressed-like manners in these interactions. Other studies have found that behavior problems and irritability in infants are associated with the persistence (Ghodsian, Zajicek,

& Wolkind, 1984) and onset of maternal clinical depression (Murray, Stanley, Hooper, King, & Fiori-Cowley, 1996).

One study addressed the methodological limitations of correlational studies of child effects research by using an experimental design. Pelham and colleagues (1997) asked married couples and single mothers to interact with 5- to 12-year-old boys who were trained to behave in either a normal or defiant manner. While waiting to have a second interaction with the same boy, the adults completed questionnaires, including one assessing depressive symptoms. Those who had interacted with the defiant boys reported significantly higher levels of depressive symptoms than those interacting with nondefiant children.

Another study used a longitudinal design to examine whether individual differences in initial child behavior and child frontal asymmetry (i.e., electroencephalogram [EEG]) were associated with changes in maternal report of depressive symptoms 1 year later (Forbes et al., 2008). Mothers whose 3- to 9-year-old children had below average affect regulation and/or right frontal EEG asymmetry reported increased depressive symptoms 1 year later.

Timing of Effects

Although the extant literature provides both theoretical and limited empirical support for a reciprocal effects model between maternal depression and child disruptive behavior, a key question remains about the timing of these effects. Specifically, it is unclear whether there are developmental periods when bidirectional or unidirectional relations are more evident than during other periods. In general, when maternal depression has been examined in relation to child outcomes, early childhood and adolescence are thought to be times when children are particularly vulnerable (Cummings & Davies, 1994). From the perspectives of physical and social maturation, both early childhood and adolescence are times of major transition in such domains as hormonal changes and social expectations, which theoretically could be made more challenging by the presence of maternal depression. For example, during infancy and early childhood, basic psychological systems are being formed, including the ability to regulate emotions and behaviors. This is also a period when primary attachment relationships are being established, with disruptions in either physiological or social systems placing children at heightened risk of later psychopathology (Beardslee, 1986). In adolescence, individuals also face multiple challenges in physiological and social domains (e.g., sexual maturity, social roles, vocational decisions, peer influences), and maternal depression has been postulated to interfere with the developmental task of achieving a healthy separation from parents and an autonomous identity (Cummings & Davies, 1994).

In addition to early childhood and early adolescence, the transition to school, though marked by less pronounced physiological (Rimm-Kaufman &

Pianta, 2000) and cognitive (Nelson, 1996) maturation, is a time of social transition for children, particularly in the area of social networks. At formal school entry, children transition from networks consisting primarily of adults to ones consisting primarily of other children (Rimm-Kaufman & Pianta, 2000). Parents typically become less familiar with the peers and adults their children spend most of their day with and have less control over their children's activities (Pianta, Cox, Taylor, & Early, 1999). Although children typically adjust to this transition in adaptive ways, for a subset of children the transition to school may serve as a catalyst for the initiation or maintenance of externalizing trajectories (Reid, 1993). As a result, middle childhood and the transition to school also may be a time of vulnerability for both mother and child effects. To date, we know of only one study that has examined reciprocal models of mothers' internalizing symptoms and children's anxious/depressed and antisocial behaviors in middle childhood (Jaffee & Poulton, 2006). The authors found support for reciprocal relationships between mothers' internalizing symptoms and girls' (but not boys') antisocial behavior as well as children's (girls' and boys') anxious/depressed behavior from age 5 to 15. When child effects were examined in middle childhood, children's (girls' and boys') anxious/depressed behavior and girls' (but not boys') antisocial behavior were related to mothers' subsequent internalizing symptoms.

In summary, the extant literature provides theoretical models and some empirical support for reciprocal effects models that postulate both parent and child effects between maternal well-being and child problem behavior, respectively. Although reciprocal models have been tested with respect to parenting practices and child problem behavior, they have been applied less often to more general parental functioning. In what follows we offer a model for adding dimensions of parent's mental health to improve understanding of the consequences of parent–child interactions. We describe two recent studies aimed at exploring the interplay between maternal depressive symptoms and child conduct problems from the toddler period through adolescence using a sample of boys followed from infancy to adolescence. In Study 1, we examined how toddler-age disruptive behavior might be associated with trajectories of maternal depression from the toddler to the school-age period and whether such trajectories of depression are then related to child reports of antisocial activity during early adolescence. Whereas Study 1 focused on how early child disruptive behavior might be related to the longitudinal course of maternal depression, Study 2 used more closely spaced reports of maternal depression and child conduct problems to examine transactional effects over time. In both cases we chose to focus on child conduct problems as the child factor because of their greater frequency among boys (Keenan & Shaw, 1997) and their critical role in coercive transactional cycles of parent–child interaction (Patterson, 1982).

Both studies used data from the Pitt Mother & Child Project, for which participating families were recruited from the Allegheny County Women, Infants, and Children (WIC) program in the Pittsburgh metropolitan area (Shaw, Gilliom, Ingoldsby, & Nagin, 2003). The sample was restricted to boys because of the larger study's primary goal to examine antecedents of antisocial behavior. Initially, 310 mothers and sons participated in the first assessment when the boys were 1.5 years old. Because WIC serves income-eligible families, the sample was predominantly low socioeconomic status (SES; i.e., at 1.5 years, mean Hollingshead status was 24.8, indicative of an unskilled working-class sample well below the poverty level; per capita family income was $241 per month and $2,892 per year) and included a diverse sample of European American (51%), African American (36%), biracial (5%), and other (Hispanic, Asian, 6%) families. The retention rate has been fairly high, with 94% of the original sample of 310 families participating in at least one of the assessments from ages 10 and 15.

Procedures for both studies covered similar assessment points, with Study 1 including observations of child disruptive behavior at 1.5 years and teacher reports of conduct problems during early adolescence, and Study 2 including parent reports of child conduct problems from ages 2 to 8 and youth reports of antisocial behavior from ages 10 to 15. Mothers and target children were seen in the laboratory and/or home when the children were 1.5, 2, 3.5, 5, 5.5, 6, 8, 10, 11, 12, and 15 years old.

STUDY 1: TODDLER PREDICTORS OF MATERNAL TRAJECTORIES OF DEPRESSION AND ADOLESCENT ANTISOCIAL BEHAVIOR

In Study 1 we used a long-term perspective to identify early forms of child disruptive behavior that might be associated with trajectories of maternal depression that we hypothesized would be related to subsequent youth and teacher reports of child antisocial behavior during adolescence. These child behaviors were negative emotionality, noncompliance, and aggression. We also examined adolescent internalizing problems as an outcome variable to address the issue of specificity of associations between maternal depression and later child adjustment. Trajectories of maternal depression were generated using a semiparametric modeling technique (Nagin, 2005) covering an 8.5-year span when children were between 1.5 and 10 years old. Child disruptive behaviors at 1.5 years were selected on the basis of their potential for affecting maternal well-being during a time when parenting efficacy has been shown to be at a low point during early childhood (Fagot & Kavanagh, 1993). Finally, completing the transactional cycle, we tested whether more persistent and severe trajectories of maternal depressive symptoms would be associ-

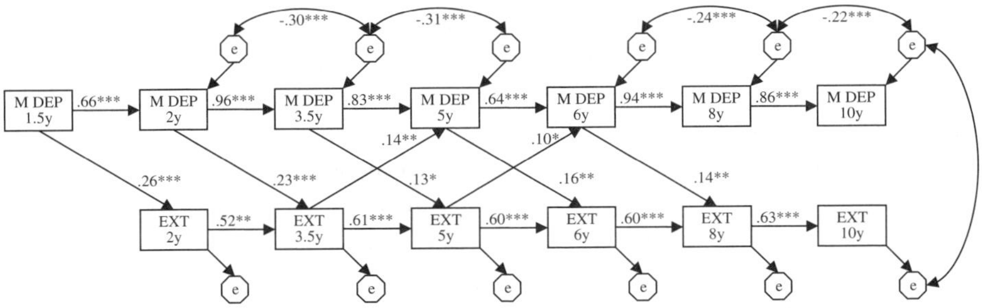

Figure 5.1. Autoregressive model for boys ages 1.5 to 10 years. M DEP = maternal depression; EXT = externalizing symptoms; e = error; y = years. * $p < .05$. ** $p < .01$. *** $p < .001$.

ated with higher levels of boys' problem behavior during early adolescence. The structure of the design is depicted in Figure 5.1.

Step 1 in the analysis was to identify trajectories of maternal depression, following procedures recommended by Nagin (2005). The best-fitting model included four groups. Groups 1 and 2, comprising the majority of the sample, demonstrated Beck Depression Inventory scores suggestive of minimal depressive symptomatology (Beck, Steer, & Garbin, 1988). Group 1, which we termed *low*, consisted of 25.2% of the sample who endorsed very few depressive symptoms across all time points, averaging about 2 across the eight time points. Group 2, which we termed *moderate low*, included 45.7% of the sample and was characterized by a consistent pattern of moderately low depressive symptoms, averaging about 6 across eight time points. Group 3 included 21.8% of mothers who were *moderate high* in number of symptoms, averaging about 12 across the time points but showing a slight decrease from 13 at age 1.5 to 11 at age 10. Group 4, which included only 7.3% of the sample, was termed the *high chronic group*, with scores suggestive of moderate to severe levels of depression at all time points averaging over 20 symptoms.

With trajectory groups of maternal depression identified, we then sought to test the first path in the model in Figure 5.1, examining associations between the different indexes of observed child disruptive behavior (i.e., aggression, noncompliance, negative emotionality) observed at age 1.5 and trajectories of maternal depression from ages 1.5 to 10. Initially, we conducted a series of analyses of variance to examine associations between trajectory group membership and individual child risk factors. Although there were significant group differences in the expected direction for each of the three types of child disruptive behavior, noncompliance showed the most consistent pattern of results, with children of mothers in the high-chronic and moderate-high depression groups found to have significantly higher noncompliance scores than children of mothers in the low group.

We then explored the second path in the model, between trajectories of maternal depression and adolescent problem behavior, using a series of analyses of covariance in which youth reports of antisocial behavior and teacher reports of externalizing and internalizing symptoms were explored with respect to maternal depression group, controlling for observed child noncompliance at 1.5 years. The pattern of results revealed that for both youth and teacher reports of antisocial/externalizing, sons of mothers in the moderate-high depression group showed significantly higher levels of problem behavior than sons of mothers in the low and moderate-low depression groups. Although means were in the expected direction, differences between the persistent-high-depression group and child externalizing failed to attain statistical significance. No significant differences were found with respect to teacher reports of internalizing for the persistent-high or moderate-high in relation to the low and moderate-low depression groups.

The results from this study were consistent with a transactional perspective of developmental psychopathology that has emphasized the dynamic interplay between child and parenting characteristics and subsequent increased risk for child maladjustment (Sameroff & Chandler, 1975). When different types of disruptive child behaviors were examined at age 1.5 within a multivariate framework controlling for family SES and income, observed child noncompliance contributed significant variance to high and persistent trajectories of the maternal depression trajectory group. In turn, moderately high trajectories of depression were related to subsequent youth reports of antisocial behavior and teacher reports of externalizing, but not internalizing, symptoms. Although there were some caveats to the consistency of the pattern of findings discussed in the section that follows (i.e., why were persistently high trajectories of maternal depression not related to boys' later antisocial behavior?), the overall pattern is consistent with Sameroff's model of development emphasizing mutually constructive dynamic interplay between children and their primary socializing agents.

STUDY 1: CHILD EFFECTS ON TRAJECTORIES OF MATERNAL DEPRESSION

In Study 1, child noncompliance was found to be the most consistent predictor of maternal trajectory group status, with child aggression and infant negative emotionality showing few or no significant effects when examined within a multivariate framework. In studies of preschool and school-age hyperactive children, many of whom exhibit high rates of noncompliance, noncompliant and oppositional behaviors in children have been found to create substantial distress for parents (Barkley, Karlsson, & Pollard, 1985). In addition, in interactions with hyperactive and noncompliant children, parents display more disapproval, are more negative and reprimanding, use more physical punishment, and are less responsive (Johnston & Mash, 2001).

Although it was not surprising to find oppositional behavior to be associated with trajectories of maternal depression, we were somewhat surprised that child aggression and negative emotionality were less consistently associated with trajectories of maternal well-being. Part of the reason for the difference may involve the greater frequency of noncompliance demonstrated by toddlers compared with rates of aggression and severe bouts of negative emotionality, a difference that has been corroborated in our laboratory- and home-based assessments with toddlers and parents. Thus, we suspect that oppositional and defiant child behaviors are likely to have a greater cumulative effect on parental mood than less frequent incidences of physical aggression or less intense bouts of negative emotionality during the second year. The timing of our assessment might have also affected the magnitude of effect for negative emotionality because previous research has shown it to be a correlate of elevated maternal depressive symptoms during the first year of life before children show increased levels of aggressive and noncompliant behavior (Murray, Stanley, et al., 1996).

STUDY 1: EFFECTS OF MATERNAL DEPRESSION ON LATER YOUTH ANTISOCIAL BEHAVIOR

Turning to the analyses that linked trajectories of maternal depression to boys' later antisocial behavior, consistent with the extant literature, we found that higher levels of maternal depressive symptoms were associated with both youth and teacher reports of antisocial outcomes at ages 11 to 13 (Zahn-Waxler, Iannotti, Cummings, & Denham, 1990). Although sons of mothers with high chronic trajectories of depressive symptoms showed somewhat higher levels of youth- and teacher-reported antisocial behavior in early adolescence, it was only the sons of mothers in the moderate-high group who consistently demonstrated significantly higher levels of antisocial outcomes when compared with boys whose mothers had low and moderate rates of depressive symptoms. Although somewhat paradoxical, this finding is consistent with a model posited by Rutter (1990), who suggested that children with chronically depressed parents may be less impaired than those with chronic but more moderate symptom levels. Accordingly, Rutter suggested that offspring of mothers who are chronically depressed are better able to understand their parents as *ill* and as a result may learn to rely more heavily on others and themselves to develop adaptive coping skills and better social functioning (Petersen et al., 1993). In contrast, when parents show a less severe, albeit a persistent and moderate course of symptoms, children may continue to expect mothers to be the primary source of support and model for developing emotion regulation skills.

Regarding child effects on maternal depressive symptoms, although we do not wish to advocate that early forms of disruptive child behavior are a primary cause of maternal depression, the findings do suggest that aversive child behaviors might merit consideration as one of a constellation of intrapersonal (e.g., rumination) and interpersonal (e.g., social support) factors previously found to affect the course of depression (e.g., Nolen-Hoeksema, 1991). Of course, without reports of maternal depression during their child's infancy or the prenatal period, it also remains unclear how early maternal depressive symptoms might have been associated with negative emotionality, noncompliance, or aggression at age 1.5. As there is no correct starting point in capturing the genesis of reciprocal effects between mothers and their children, this study should be considered only a representative slice of a transactional process. Nonetheless, as previous literature has placed little emphasis on how child factors might have an impact on the course of maternal depression after the 1st year of life (Murray, Stanley, et al., 1996), the findings suggest that child disruptive behavior might be an informative risk factor to assess among mothers showing high levels of depressive symptoms. It is somewhat surprising that child effects are not considered more heavily in assessing risk of maternal depression for families with toddler-age children on the basis of the high levels of oppositional and defiant behaviors exhibited during the "terrible twos" (Shaw & Bell, 1993).

STUDY 2: TRANSACTIONAL EFFECTS BETWEEN MATERNAL DEPRESSION AND CHILD CONDUCT PROBLEMS AND ANTISOCIAL BEHAVIOR FROM TODDLERHOOD TO ADOLESCENCE

In Study 1, we used more of a macro approach to examine how early child disruptive behavior may be related to trajectories of maternal depression and, in turn, subsequent adolescent antisocial behavior. In contrast, Study 2 was designed to more explicitly examine the validity of Sameroff's original transactional model, exploring parent and child effects of maternal depression and child conduct problems between ages 1.5 and 8 and using a youth report measure of antisocial activities to examine similar transactional processes between ages 10 and 15. For the earlier age period, the model controls for autoregressive effects of maternal depression and child externalizing at each age and simultaneously addresses how much variance is contributed across variables to the next assessment of maternal depression and child conduct problems. Latent growth curve modeling (LGCM) and structural equation modeling were used to carry out these analyses. Initially, LGCM was used to identify whether there were changes in maternal depression or boys' conduct problems over time for both age periods (Bryk & Raudenbush, 1987). We then investigated whether sons' externalizing and mothers' depression were related to one another over time, generating cross-lagged associations between maternal depression and boys' externalizing in an autoregressive path model, controlling for stability of each construct by simultaneously regressing temporally later scores on earlier ones (e.g., maternal depressive symptoms at age 10 were regressed on child externalizing symptoms at age 8). Significant cross-lagged paths indicate the timing of child effects on maternal symptoms, and maternal effects on child problem behaviors.

Model 1: Ages 1.5 to 10

Initially, the best fitting model for growth in maternal depression from child ages 1.5 to 10 was found to be quadratic, with mothers initially reporting moderately high levels of depression, followed by gradual declines by age 10. For boys' growth of conduct problems, a cubic model was found to have the best fit. On average, boys' initially high levels of externalizing problems at age 2 increased further by age 3.5, then subsequently declined by age 10.

Next, an autoregressive path model was fitted to examine the potential for child effects on maternal depression and of maternal effects on sons' conduct problems. In this model, which controls for the stability of each construct, we examined the cross-lagged associations between maternal depression and child conduct problems. As displayed in Figure 5.1, the results indicated that five of the six paths from maternal depression to boys' subsequent conduct problems were significant. Maternal levels of depression at ages 1.5, 2, 3.5, 5, and 6 years were positively related to higher levels of boys' later conduct problems at ages 2, 3.5, 5, 6, and 8 years, respectively. In contrast, only two of six paths from boys' conduct problems to maternal depression were significant; boys' conduct problems at ages 3.5 and 5 were positively related to maternal depression at ages 5 and 6, respectively.

Model 2: Ages 10 to 15

The same procedures were then used to examine growth curves and autoregressive models between maternal depression and boys' reports of antisocial behavior between ages 10 and 15. The best fit of growth in maternal depression during this period was a linear model, with a modest positive, albeit nonsignificant, slope between ages 10 and 15. For boys' antisocial behavior, a quadratic function provided the best fit; boys' average initial antisocial behavior was low, remained relatively stable through age 12, and then increased markedly between ages 12 and 15.

Results of the autoregressive model during the early adolescent period were similar to those found in early and middle childhood. As shown in Figure 5.2, two of the three paths from maternal depression to boys' antisocial behavior were significant predictors such that higher levels of maternal depression at ages 11 and 12 were predictive of boys' later self-reported antisocial

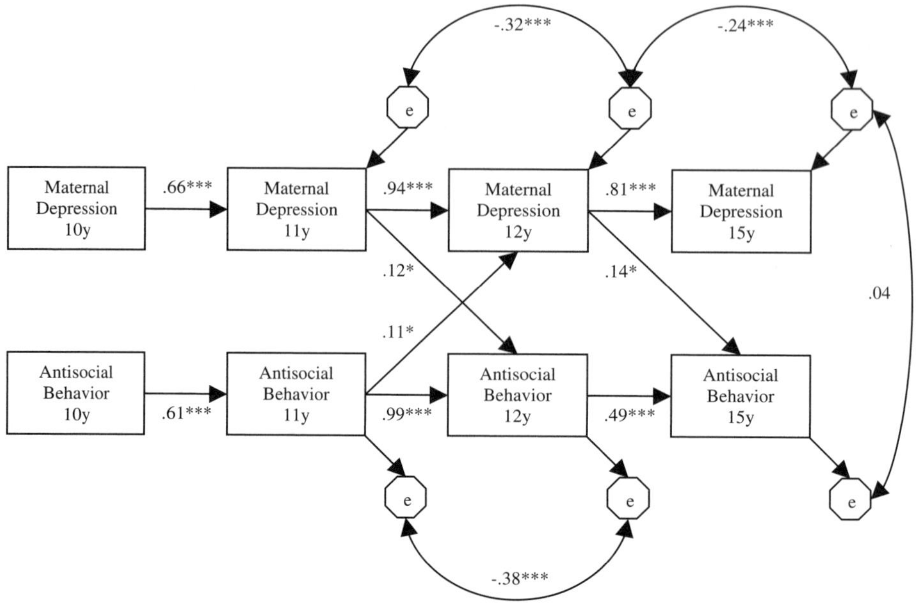

Figure 5.2. Autoregressive model (replication from boys ages 10 to 15 years). e = error; y = years. * $p < .05$. ** $p < .01$. *** $p < .001$.

behavior at ages 12 and 15, respectively. In comparison, only one of three paths from boys' antisocial behavior to maternal depression was significant; higher levels of boys' antisocial behavior at age 11 were associated with higher levels of maternal depression at age 12.

These results were less consistent, albeit not absent, in demonstrating that child conduct problems in early childhood and antisocial activities in early adolescence were related to crossing the intrapersonal barrier and affecting later maternal depressive symptoms. In contrast and in accord with previous research, current levels of maternal depression were a consistent predictor of subsequent child conduct problems (Beardslee et al., 1998). Although the percentage of significant associations between maternal depression and later child behavior might have been inflated because of reporter bias in the early-middle childhood model (i.e., five of six correlations were significant) when mothers reported on both their own depressive state and child conduct problems (Fergusson et al., 1993), a similar percentage (i.e., two of three) of significant cross-lagged correlations emerged during early adolescence when youth reported on their antisocial activities. In contrast, in only 33% of cases during early-middle childhood and early adolescence (i.e., two of six and one of three instances, respectively) were child effects evident in relation to the next assessment of maternal depression using the same and different reporters of child disruptive and antisocial behavior.

STUDY 2: TIMING OF RECIPROCAL EFFECTS OF MATERNAL DEPRESSION AND CHILD CONDUCT PROBLEMS

In Study 2, we expected that parent and child effects would be more prominent in periods of physical and social transition than was supported by the data. Maternal depression was most strongly related to child externalizing behavior in toddlerhood. Parent effects were still significant, although slightly weaker, during the transition to middle childhood. Finally, during the transition to early adolescence, associations between maternal depression and boys' delinquent behavior were once again evident.

The current findings were generally consistent with the premise that child effects would be found during times of physical and social transition for children, but there were exceptions to this expected trend. For instance, consistent with the results from Study 1, we expected to see child effects on maternal depression during the terrible twos, but such associations were not evident until later in the preschool period (i.e., age 3.5). The discrepancy in findings might be at least partially a result of the measurement of different behaviors (narrow vs. broader indexes of disruptive behavior), the use of different informants (i.e., observers vs. mothers) and different periods of time to assess child disruptive behavior (i.e., several minutes vs. several months) in the two studies. Future studies examining child effects on maternal well-being may want to focus on a narrower range of child disruptive behaviors that are aversive, frequently occurring (i.e., noncompliance is the hallmark of the terrible twos), and possibly more sensitive to mothers prone to experience elevated rates of depressive symptoms.

DISCUSSION

In describing the process of raising a son by herself, a mother participating in Mavis Hetherington's study on divorce and children's adjustment described the process akin to "being pecked to death slowly by ducks" (M. Hetherington, personal communication, October 1984). Indeed, much of the research on coercive processes was developed on the basis of anecdotal and scientific reports of noxious interactions male children have with their mothers (Patterson, 1982). Similarly, theories on the determinants of parenting have postulated that the quality of caregiving practices can be compromised by individual differences in child problem behavior and maternal well-being (Shaw & Bell, 1993). Our work sought to expand our understanding of dyadic processes between mothers and sons by applying a bidirectional perspective to the issue of maternal depression and child disruptive behavior from early childhood through adolescence. In essence, we sought to address the question of whether having a disruptive child affected internal states of maternal well-being, and in turn whether levels of maternal depression were associated with later indexes of child disruptive and antisocial behavior. In accord with Bugental's work (see chap. 6, this volume), we are fairly confident that maternal depression bodes poorly for parenting competence and in early childhood is associated with risk of conduct problems in middle childhood (Shaw, Bell, & Gilliom, 2000). The question then becomes whether living with a disruptive child appears to cross the intrapersonal boundary of being more than an irritant for mothers and actually is related to future levels of depressive symptoms. In accord with reciprocal (Bell, 1968) and transactional (Sameroff, 1995) models, we designed these two studies with the expectation that child effects would prove to be consistent predictors of later maternal depression. Some of the results from both studies were consistent with our hypothesis; however, this was not uniformly the case, particularly in Study 2, which represented a more proximal test of the original transactional model.

Broader Issues and Future Directions

Although results from Study 1 indicate support for transactional effects across broad periods of time and results from Study 2 provide a more microlevel perspective across 1- to 2-year periods, the current results do not address specific mechanisms by which symptoms of maternal depression and child conduct problems might affect one another. Such mechanisms may be evident by examining moment-by-moment observations of parent–child behavior (see chap. 2, this volume). A number of investigators have suggested and found support for the notion that associations between parental depression and child adjustment are mediated by parenting, specifically tendencies for depressed mothers to be negative, critical, unresponsive, helpless, and low on positivity toward offspring (Goodman & Gotlib, 1999). More intensive examinations of this process also provide some support for this notion. For example, Shaw, Schonberg, et al. (2006) explored mothers' contingent responses to their children's expression of emotions. They found that mothers with a history of childhood-onset depression (COD) showed less contingent responsivity to their child's expression of sadness and distress than non-COD mothers. Additional research is needed to more intensively uncover the processes by which parent-to-child and child-to-parent effects influence child conduct problems and maternal well-being, respectively, as well as child (see chap. 7, this volume) and maternal (see chap. 6, this volume) interpretations and representations of their partner's behavior.

In addition, as children spend more and more time outside of the home interacting with peers and nonrelated adults and relatively less time interacting with parents with the onset of formal schooling, it is logical that transactional models target changes of emphasis in children's developmental stepladder of socialization by focusing on interactions children have with peers and other important adults in their daily lives. Dishion, McCord, and Poulin's (1999) work on peer contagion in relation to deviancy training with adolescents and Snyder et al.'s (2008) recent extension of this work to school-age children using playground observations provide examples of extending a transactional perspective to developmentally critical phenomena. Future research on school-age children and adolescents would benefit from incorporating the study of interpersonal exchanges with peers (e.g., reinforcement of deviant behavior) into children's intrapersonal representations and from observing these reciprocal processes unfold longitudinally (e.g., examining the effects of social information processing on later deviancy training and exploring the effects of deviancy training on later social information processing).

Another question raised by these findings is whether there are other child behaviors in addition to child externalizing symptoms that are related to increases in maternal depression. For example, Jaffee and Poulton (2006) found boys' symptoms of anxiety and depression to be related to maternal internalizing symptoms, rather than externalizing problems. Other unexplored child characteristics that may affect maternal well-being and warrant investigation include medical or pervasive developmental disorders (e.g., mental retardation, autism), and child academic and peer-related difficulties.

Limitations

There are a few significant methodological limitations to both studies. First, participants were primarily low-income European American and African American boys living in an urban setting. Some research suggests that there are gender differences in how children are affected by maternal depression. For example, Leve, Kim, and Pears (2005) found that the association between maternal depressive symptoms and later outcomes varied by gender, with elevated maternal depressive symptoms uniquely predicting increases in internalizing symptoms for girls and increases in boys' externalizing symptoms when boys' impulsivity was low. Similarly, child effects on maternal depression may vary by gender. Steinberg (2001) found that in adolescence, parental distress from parent–child conflict was more intense for parents whose adolescent is the same sex. Jaffee and Poulton (2006) found that in middle childhood and early adolescence, girls' antisocial behavior predicted increases in maternal depression, but boys' antisocial behavior did not. Moreover, it is not clear whether these results would have been replicated for children living in rural or suburban contexts. Finally, one cannot rule out the role of possible third variables at age 1.5 or during the period from ages 1.5 to 10 when maternal depressive symptoms were measured. Incorporating measurement of other child (e.g., inhibitory control) or parent (e.g., caregiving quality) factors or extrafamilial socializing agents (e.g., teachers at school, peers in the neighborhood or at school) might provide additional insight into the underlying mechanisms by which reciprocal and transactional associations emerge.

Clinical Implications

The results of both studies are consistent with the robust finding in the existing literature that higher levels of maternal depressive symptoms are associated with poor outcomes in children. This finding, coupled with our finding that maternal depressive symptoms were relatively stable over time and related to more serious forms of antisocial behavior in adolescence, indicates the need for early identification and preventive interventions (Olds, 2002; Shaw, Dishion, Supplee, Gardner, & Arnds, 2006). In addition, the significant child effects of externalizing problems on increases in maternal depressive symptoms during the terrible twos in Study 1 and during the transition to middle childhood (ages 3.5 to 5) and adolescence (ages 11 to 12) in Study 2 suggest that mental health clinicians working with depressed mothers include an assessment of child behavior and its impact on maternal well-being for mothers of youth approaching school-age and adolescence.

In sum, these findings provide novel information about the transactional processes between child conduct problems and maternal depression examining both relatively short- and long-term bidirectional effects spanning from early childhood through early adolescence. The results shed new light on transactional processes in a sample of mothers who showed above average rates of depressive symptoms and boys who exhibited elevated rates of problem behaviors throughout childhood and adolescence.

REFERENCES

Baker, B. L., & Heller, T. L. (1996). Preschool children with externalizing behaviors: Experience of fathers and mothers. *Journal of Abnormal Child Psychology, 24,* 513–532.

Barkley, R. A., Karlsson, J., & Pollard, S. (1985). Effects of age on the mother–child interactions of ADD-H and normal boys. *Journal of Abnormal Child Psychology, 13,* 631–637.

Beardslee, W. R. (1986). The need for the study of adaptation in the children of parents with affective disorders. In M. Rutter, C. E. Izard, & P. B. Read (Eds.), *Depression in young people* (pp. 189–204). New York: Guilford Press.

Beardslee, W. R., Versage, E. M., & Gladstone, T. G. (1998). Children of affectively ill parents: A review of the past 10 years. *Journal of the American Academy of Child & Adolescent Psychiatry, 37,* 1134–1141.

Beck, A. T., Steer, R. A., & Garbin, M. G. (1988). Psychometric properties of the Beck Depression Inventory: Twenty-five years of evaluation. *Clinical Psychology Review, 8,* 77–100.

Bell, R. Q. (1968). A reinterpretation of the direction of effects in studies of socialization. *Psychological Review, 75*, 81–95.

Bell, R. Q., & Harper, L. V. (1977). *Child effects on adults*. Hillsdale, NJ: Erlbaum.

Belsky, J. (1984). The determinants of parenting: A process model. *Child Development, 55*, 83–96.

Bryk, A. S., & Raudenbush, S. W. (1987). Application of hierarchical linear models to assessing change. *Psychological Bulletin, 101*, 147–158.

Campbell, S. B., Brownell, C. A., Hungerford, A., Spieker, S., Mohan, R., & Blessing, J. S. (2004). The course of maternal depressive symptoms and maternal sensitivity as predictors of attachment security at 36 months. *Development and Psychopathology, 16*, 231–252.

Campbell, S. B., Shaw, D. S., & Gilliom, M. (2000). Early externalizing behavior problems: Toddlers and preschoolers at risk for later maladjustment. *Development and Psychopathology, 12*, 467–488.

Civic, D., & Holt, V. L. (2000). Maternal depressive symptoms and child behavior problems in a nationally representative normal birth weight sample. *Maternal and Child Health Journal, 4*, 215–221.

Conger, R. D., Patterson, G. R., & Ge, X. (1995). It takes two to replicate: A mediational model for the impact of parents' stress on adolescent adjustment. *Child Development, 66*, 80–97.

Coyne, J., Kahn, J., & Gotlib, I. (1987). Depression. In T. Jacob (Ed.), *Family interaction and psychopathology: Theories, methods, and findings* (pp. 509–533). New York: Plenum Press.

Cummings, E. M., & Davies, P. T. (1994). Maternal depression and child development. *Journal of Child Psychology and Psychiatry, 35*, 73–112.

Cummings, E. M., Keller, P. S., & Davies, P. T. (2005). Towards a family process model of maternal and paternal depressive symptoms: Exploring multiple relations with child and family functioning. *Journal of Child Psychology and Psychiatry, 46*, 479–489.

Dishion, T. J., McCord, J., & Poulin, F. (1999). When interventions harm: Peer groups and problem behavior. *American Psychologist, 54*, 755–764.

Donenberg, G., & Baker, B. L. (1993). The impact of young children with externalizing behaviors on their families. *Journal of Abnormal Child Psychology, 21*, 179–186.

Eddy, J. M., Leve, L. D., & Fagot, B. I. (2001). Coercive family processes: A replication and extension of Patterson's coercion model. *Aggressive Behavior, 27*, 14–25.

Elgar, F. J., McGrath, P. J., Waschbusch, D. A., Stewart, S. H., & Curtis, L. J. (2004). Mutual influences on maternal depression and child adjustment problems. *Clinical Psychology Review, 24*, 441–459.

Fagot, B. I., & Kavanagh, K. (1993). Parenting during the second year: Effects of children's age, sex, and attachment classification. *Child Development, 64*, 258–271.

Fergusson, D. M., Lynskey, M. T., & Horwood, L. J. (1993). The effect of maternal depression on maternal ratings of child behavior. *Journal of Abnormal Child Psychology, 21*, 245–269.

Field, T., Healy, B., Goldstein, S., Perry, S., Bendell, D., Shanberg, S., et al. (1988). Infants of depressed mothers show "depressed" behavior even with nondepressed adults. *Child Development, 59*, 1569–1579.

Forbes, E. E., Shaw, D. S., Silk, J. S., Feng, X. I., Cohn, J. F., Fox, N. A., & Kovacs, M. (2008). Children's affect expression and frontal EEG asymmetry: Transactional associations with mothers' depressive symptoms. *Journal of Abnormal Child Psychology, 36*, 207–221.

Ghodsian, M., Zajicek, E., & Wolkind, S. (1984). A longitudinal study of maternal depression and child behaviour problems. *Journal of Child Psychology and Psychiatry and Allied Disciplines, 25*, 91–109.

Goodman, S. H., & Brumley, H. E. (1990). Schizophrenic and depressed mothers: Relational deficits in parenting. *Developmental Psychology, 26*, 31–39.

Goodman, S. J., & Gotlib, I. H. (1999). Risk for psychopathology in the children of depressed mothers: A developmental model for understanding mechanisms of transmission. *Psychological Review, 106*, 458–490.

Hay, D. F., Pawlby, S., Angold, A., Harold, G. T., & Sharp, D. (2003). Pathways to violence in the children of mothers who were depressed postpartum. *Developmental Psychology, 39*, 1083–1094.

Jaffee, S. R., & Poulton, R. (2006). Reciprocal effects of mothers' depression and children's problem behaviors from middle childhood to early adolescence. In A. C. Huston, M. N. Ripke, & J. McCord (Eds.), *Developmental contexts in middle childhood: Bridges to adolescence and adulthood* (pp. 107–129). New York: Cambridge University Press.

Johnston, C., & Mash, E. J. (2001). Families of children with attention-deficit/hyperactivity disorder: Review and recommendations for future research. *Clinical Child and Family Psychology Review, 4*, 183–207.

Keenan, K., & Shaw, D. S. (1997). Developmental influences on young girls' behavioral and emotional problems. *Psychological Bulletin, 121*, 95–113.

Leve, L. D., Kim, H. K., & Pears, K. C. (2005). Childhood temperament and family environment as predictors of internalizing and externalizing trajectories from ages 5 to 17. *Journal of Abnormal Child Psychology, 33*, 505–520.

Leve, L. D., Scaramella, L. V., & Fagot, B. I. (2001). Infant temperament, pleasure in parenting, and marital happiness in adoptive families. *Infant Mental Health Journal, 22*, 545–558.

Marchand, J. F., Hock, E., & Widaman, K. (2002). Mutual relations between mothers' depressive symptoms and hostile-controlling behavior and young children's externalizing and internalizing behavior problems. *Parenting: Science and Practice, 2*, 335–353.

Murray, L., Fiori-Cowley, A., Hooper, R. & Cooper, P. J. (1996). The impact of postnatal depression and associated adversity on early mother infant interactions and later infant outcome. *Child Development, 67*, 2512–2516.

Murray, L., Stanley, C., Hooper, R., King, F. & Fiori-Cowley, A. (1996). The role of infant factors in postnatal depression and mother–infant interactions. *Developmental Medicine and Child Neurology, 38*, 109–119.

Nagin, D. S. (2005). *Group-based modeling of development*. Cambridge, MA: Harvard University Press.

Nelson, D. R., Hammen, C., Brennan, P. A., & Ullman, J. B. (2003). The impact of maternal depression on adolescent adjustment: The role of expressed emotion. *Journal of Counseling and Clinical Psychology, 71*, 935–944.

Nelson, K. (1996). Memory development from 4 to 7 years. In A. J. Sameroff & M. M. Haith (Eds.), *The five to seven year shift: The age of reason and responsibility* (pp. 141–160). Chicago: The University of Chicago Press.

Nolen-Hoeksema, S. (1991). Responses to depression and their effects on the duration of depressive episodes. *Journal of Abnormal Psychology, 100*, 569–582.

Olds, D. (2002). Prenatal and infancy home visiting by nurses: From randomized trials to community replication. *Prevention Science, 3*, 153–172.

Patterson, G. R. (1982). *Coercive family processes*. Eugene, OR: Castalia.

Pelham, W. E., Lang, A. R., Atkeson, B., Murphy, D. A., Gnagy, E. M., Greiner, A. R., et al. (1997). Effects of deviant child behavior on parental distress and alcohol consumption in laboratory interactions. *Journal of Abnormal Child Psychology, 25*, 413–424.

Petersen, A. C., Compas, B. E., Brooks-Gunn, J., Stemmler, M., Ey, S., & Grant, K. E. (1993). Depression in adolescence. *American Psychologist, 48*, 155–168.

Pianta, R. C., Cox, M. J., Taylor, L., & Early, D. (1999). Kindergarten teachers' practices related to the transition to school: Results of a national survey. *The Elementary School Journal, 100*, 71–86.

Reid, J. B. (1993). Prevention of conduct disorder before and after school entry: Relating interventions to developmental findings. *Development and Psychopathology, 5*, 243–262.

Rimm-Kaufman, S. E., & Pianta, R. C. (2000). An ecological perspective on the transition to kindergarten: A theoretical framework to guide empirical research. *Journal of Applied Developmental Psychology, 21*, 491–511.

Rutter, M. (1990). Commentary: Some focus and process considerations regarding effects of parental depression on children. *Developmental Psychology, 26*, 60–67.

Sameroff, A. J. (1995). General systems theories and developmental psychopathology. In D. Cichetti & D. Cohen (Eds.), *Developmental psychopathology: Theory and methods* (Vol. 1, pp. 659–695). New York: Wiley.

Sameroff, A. J., & Chandler, M. (1975) Reproductive risk and the continuum of caretaking casualty. In F. D. Horowitz, E. M. Hetherington, S. Scarr-Salapatek & G. Siegel (Eds.), *Review of child development research* (Vol. 4, pp. 187–204). Chicago: The University of Chicago Press.

Shaw, D. S., & Bell, R. Q. (1993). Developmental theories of parental contributors to antisocial behavior. *Journal of Abnormal Child Psychology, 21*, 493–518.

Shaw, D. S., Bell, R. Q., & Gilliom, M. (2000). A truly early starter model of antisocial behavior revisited. *Clinical Child and Family Psychology Review, 3*, 155–172.

Shaw, D. S., Dishion, T. J., Supplee, L., Gardner, F., & Arnds, K. (2006). Randomized trial of a family-centered approach to the prevention of early conduct problems: 2-year effects of the family check-up in early childhood. *Journal of Consulting and Clinical Psychology*, *74*, 1–9.

Shaw, D. S., Gilliom, M., Ingoldsby, E. M., & Nagin, D (2003). Trajectories leading to school-age conduct problems. *Developmental Psychology*, *39*, 189–200.

Shaw, D. S., Schonberg, M., Sherrill, J., Huffman, D., Lukon, J., Obrosky, D., & Kovacs, M. (2006). Responsivity to offspring's expression of emotion among childhood-onset depressed mothers. *Journal of Clinical Child and Adolescent Psychiatry*, *35*, 490–503.

Sinclair, D. A., & Murray, L. (1998). Effects of postnatal depression on children's adjustment in school. *British Journal of Psychiatry*, *172*, 58–63.

Snyder, J., Schrepferman, L., McEachern, A., Barner, S., Johnson, K., & Provines J. (2008). Peer deviancy training and peer coercion: Dual processes associated with early-onset conduct problems. *Child Development*, *79*, 252–268.

Steinberg, L. (2001). We know some things: Parent–adolescent relationships in retrospect and prospect. *Journal of Research on Adolescence*, *11*, 1–19.

Teti, D. M., & Gelfand, D. M. (1991). Behavioral competence among mothers and infants in the first year: The mediational role of maternal self-efficacy. *Child Development*, *62*, 918–929.

Whiffen, V. E., & Gotlib, I. H. (1989). Infants of postpartum depressed mothers: Temperament and cognitive status. *Journal of Abnormal Psychology*, *98*, 274–279.

Zahn-Waxler, C., Iannotti, R. J., Cummings, E. M., & Denham, S. (1990). Antecedents of problem behaviors in children of depressed mothers. *Development and Psychopathology*, *2*, 271–293.

6

PREDICTING AND PREVENTING CHILD MALTREATMENT: A BIOCOGNITIVE TRANSACTIONAL APPROACH

DAPHNE BUGENTAL

In this chapter, I consider the pathology-amplifying transactional processes that involve (a) adults who interpret caregiving challenges as threats and who respond to such threats with defensive reactions and (b) children who are exceptionally reactive to their social environment. Within this dyadic combination, the responses shown by one member foster increased stress and maladaptive coping responses in the other member (see Figure 6.1). This process reflects a biocognitive transactional system that may serve to foster maltreatment or mistreatment within parent–child relationships and may even lead to the perpetuation of such interactions across generations (e.g., Francis, Diorio, Liu, & Meaney, 1999; Maestripieri, 2005).

I also consider the utility of intervention processes early in life that may prevent or attenuate the mistreatment or maltreatment of young children that often follows from combined parent and child risk features. A transactional approach can be used not only to explain or predict child maltreatment but also as a conceptual framework for remediation.

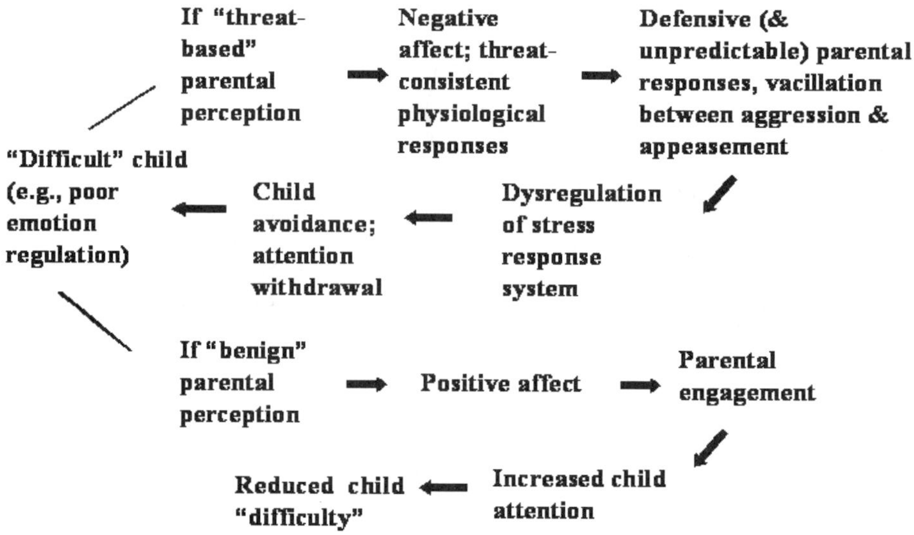

Figure 6.1. Biocognitive transactional model of child maltreatment.

BACKGROUND OF THE QUESTION

Early approaches to child maltreatment at both a theoretical level and a popular level accounted for such processes with simple linear models. Psychologists typically located causality within parental psychopathology because it was believed that only parents with some type of mental illness would engage in such behavior. Sociologists, in contrast, were more likely to propose a linear model in which causality resided within stressors and problems that occur in society. Those parents who were more likely to engage in harsh or abusive practices often located the source of problems within the misdeeds and provocations of the children. Reactions across time and within different cultures also varied. For example, infanticide in traditional societies is often viewed as appropriate when such children were seen as evil or as hopelessly flawed (see Bugental, 2003, for a review of cultural differences). So the story begins with competing but consistently linear accounts of the causes of child maltreatment.

My own interest in this topic emerged at a time when psychologists were first questioning linear accounts of maltreatment or indeed linear accounts of parenting and socialization processes in general. As a social psychologist, I was employed in a postdoctoral position with the mission of exploring the communication processes that occur in dysfunctional (and often abusive) families in comparison with typical families. My training led me to focus on the kinds of attributions that parents in dysfunctional families were making about their children. Not only were parents blaming children for their negative experiences at school and at home but they were also blaming children for the problems created for them as parents. In short, such parents appeared to think of themselves as victims of their children. At the same time, the children—even from the perspective of a objective observer—could best be described as "difficult," "odd," or "hard to handle." Although I found systematic differences in the communication style of dysfunctional versus typical families, these observations did not consider the dynamic source of these differences. These findings (and anecdotal observations) did, however, serve as a motivation to explore the current state of theory and knowledge on this topic.

Emerging Models of Child Maltreatment

In exploring past accounts of poorly functioning or maltreating families, it was easy to see the focus on the characteristics and effects of parents (Spinetta & Rigler, 1972: Steele & Pollack, 1968). This approach was consistent with the then-prevailing parent effects model. However, there was also emerging evidence that children's characteristics influenced their risk of maltreatment (e.g., de Lissovoy 1979; Elmer & Gregg, 1967; Sherrod, O'Connor, Vietze, & Altemeier, 1984), consistent with the general challenge offered by Bell (1968) to parent effects models.

Transactional approaches were introduced to address the reciprocal influences between parents and offspring, that is, the changing processes that occur over time within dyadic (or family) relationships (as originally proposed by Sameroff, 1975; Sameroff & Chandler, 1975; and as reviewed later by Sameroff & MacKenzie, 2003). The notion of reciprocal influences within maltreating relationships was subsequently reflected in the social contextual model of parenting described by Belsky (1984).

When attention was directed to pathological processes, attention was also given to the particular dyadic combinations that promote or attenuate psychopathology. This basic idea was a component of Thomas, Chess, and Birch's (1968) focus on the match (or mismatch) between parents and children. These investigators suggested that children with a difficult temperament would show quite different outcomes based on the ways in which parents responded to these child characteristics. Relevant to pathological processes, negative outcomes were predicted when parents were negatively reactive to a child's temperament pattern. When focused on developmental psychopathology, transactional models were extended to conceptualize such outcomes as residing in the adaptiveness (or maladaptiveness) of the relationship between individual and context (Sameroff & Emde, 1989).

Development of Cognitive Model of Child Maltreatment

Following my training in social cognition, I worked in a lab where attention was turned to the explanatory processes that appeared to foster negative reactivity to certain kinds of children, that is children who might be construed as resistant or unresponsive. Social cognition theory drew attention to the fact that people are more likely to rely on their preexisting beliefs and schemas when faced with an ambiguous situation, in particular one that posed a potential problem or challenge (Wong & Weiner, 1981).

The early observations within this program of research, which my colleagues and I conducted, drew attention to (a) the distinctive features of children who were at greater risk of harsh or abusive parenting, along with (b) the apparent explanatory biases of their parents. The families studied had been referred to clinical settings as a result of the problems parents experienced in coping with difficult caregiving experiences (Bugental, Blue, & Cruzcosa, 1989; Bugental, Blue, & Lewis, 1990).

The children who were more likely to be identified as difficult (by both parents and objective observers) and who were the recipients of harsher or more abusive parenting showed different behavioral patterns than did their easier (and less harshly treated) siblings (Bugental et al., 1990). For example, they were less socially responsive to the initiations of either their own mothers or those of an unrelated woman. In addition, they demonstrated a variety of patterns that judges (blind to their history) interpreted as "inappropriate" or "atypical" (e.g., stereotypical movements). These children were also more likely to be reported by their mothers as having been born prematurely or as having experienced birth complications—an anecdotal account that we followed up later with a systematic prospective analysis.

A comparison was also made of the caregiving attributions made by parents who were abusive or nonabusive with their children within a clinical sample of those who shared the history of experiencing difficulty in caregiving relationships but who differed in their parenting tactics (Bugental et al., 1989). Parents who used harsh or abusive tactics were more likely than other parents to explain problematic interactions with children as a result of factors that the child could control (e.g., the child being stubborn) or of factors that the adult could not control (e.g., the adult being sick). Harsher parents were more likely to characterize themselves as experiencing a low "balance of power" in the relationship; that is, they saw the child as having more control over relationship outcomes than they did. Unrelated mothers with a low perceived balance of power were more likely to respond to the difficult child (i.e., the child more likely to receive harsh parenting at home) with more annoyance or irritation than they did to the easier child (i.e., a sibling who was less likely to receive harsh parenting at home). In addition, unrelated mothers with low perceived power responded to the more difficult child with increasing expression of negative affect over the course of the interaction (Bugental et al., 1990).

Although these early findings revealed the potential importance of parental power perception, there were limitations in the conclusions that could be drawn. For example, no information was available regarding the specific features of children that served to elicit more negative affect from unrelated women with low perceived power. In addition, the correlation between the attribution biases of mothers and the characteristics of their children allowed no secure causal inferences, in that the direction of effects was unclear. There was also the possibility that effects could be accounted for by shared genetic influences. With all of this in mind, we shifted to an experimental approach in testing the suggested transactional relationship.

EXPERIMENTAL TEST OF TRANSACTIONAL MODEL

We conducted a series of studies in which we provided an experimental test of the proposed transactional relationships. We began with a consideration of the reactions of caregivers to the stimulus posed by a child (as experimentally varied). From there we turned our attention to some of the biological mediators of the observed patterns. Finally, we explored the reactions of children to the different responses displayed by caregivers.

Adult Cognitions as Moderators of Their Reactions to Children

My colleagues and I conducted a series of investigations that allowed the possibility of drawing causal inferences in the relationships that emerge between adults and children. The research strategy involved observations of the interactions between adults who differed in their caregiving perceptions and unrelated children who differed systematically in the potential challenge or threat they posed. We proposed that the beliefs or interpretive biases that adults bring to the relationship constitute a central moderator variable in the responses shown to difficult children (Bugental, 1992; Bugental & Shennum, 1984). The settings and tasks used were relatively ambiguous in nature—a context that is more likely to lead adults to rely on their preexisting interpretive biases.

Adult participants were selected on the basis of their power perceptions (as measured by the Parent Attribution Test; Bugental et al., 1989). One line of research within this program explored the dynamic functioning of these *power schemas*. Parents with low power schemas manifested exceptionally easy access to power ideation (Bugental, Lyon, Cortez, & Krantz, 1997). Their decisions about power-relevant information were made just as quickly in the presence or the absence of a cognitive load. In contrast, other parents, when making power-relevant decisions, showed a slowed response. The short response latency shown by low-power parents in the presence of a cognitive load suggests the implicit or automatic nature of their interpretations of power-relevant information. As an indication of the conflict they experienced, they rated themselves as more dominant in the absence of a cognitive load (a condition that allows the operation of more deliberative, controlled judgments) but rated their child as more dominant in the presence of a cognitive load (a condition that serves to elicit easily accessed automatic responses). This suggests a combination of a strong motivation to maintain control (in the caregiving relationship) but a questioning of their ability to do so. Any situation that poses a potential power threat also appears to interfere with effortful thought for such individuals. For example, they had more difficulty recalling their thoughts after interacting with a child who posed a potential power threat (Bugental, Brown, & Reiss, 1996).

Within this line of research, the adult participants included both parents and nonparents (with no differences found between groupings) and both men and women (with findings generally stronger for women). The tasks involved either (a) a game format in which the adult served as an instructor (Bugental, Caporael, & Shennum 1981; Bugental et al., 1993; Bugental, Lewis, Lin, Lyon, & Kopeikin, 1999; Lewis, Bugental & Fleck, 1991), (b) an unstructured conversation in which the adult was directed to take the lead (Bugental & Shennum, 1984), (c) a situation in which the adult simply viewed (or heard) the child he or she expected to interact with or instruct (Bugental & Cortez, 1988; Lin, Bugental, Turek, Martorell, & Olster, 2002), or (d) priming for thoughts about a difficult child (Bugental et al., 1996).

Child characteristics were either selected (e.g., Bugental et al., 1981; Lin et al., 2002) or systematically varied (e.g., Bugental et al., 1993; Bugental & Lewis, 1998, 1999; Bugental, Lewis, et al., 1999; Lewis et al., 1991). Thus, in some cases, children differed naturally in their characteristics (e.g., difficult or easy in their temperament style). In other instances, children served as experimental confederates and enacted the role of a difficult or easy child. In still other instances, the behavior of the child involved a fully controlled computer simulation of a child's behavior. The outcome variables in these studies focused on adults' communication patterns (both verbal and nonverbal), affective responses, and cognitive processing.

The central findings obtained were replicated across studies. The most negative and most ambivalent reactions were shown by adults with low perceived power paired with a difficult child. Within such pairings, adults with low perceived power were more likely to show a communication pattern that revealed both their control efforts and their perceived lack of control. Their facial affect often revealed a high presence of *non-Duchenne* smiles (Bugental et al., 1996; Bugental, Kopeikin, & Lazowski, 1991), that is, smiles of a type shown when people are motivated to appear pleasant but do not actually experience positive affect. In addition, their vocal patterns often showed high levels of uncertainty (e.g., nonfluencies in speech, low vocal amplitude), combined with assertive verbal content (Bugental & Lewis, 1998, 1999). Adults with low perceived power (in comparison with those with higher levels of perceived power) often showed opposite patterns in different contexts.

In terms of their behaviors, caregivers with low perceived power—after being primed for competitive ideation—have been observed to manifest a harsher interaction style. In particular, fathers with low perceived power—primed for thoughts of competition—were more likely (than other groupings of fathers) to verbally derogate the performance either of their own child or that of an unrelated child (Bugental & Happaney, 2000). In addition, women who believed they were interacting with a child over whom they had low control (actually a computer simulation) subsequently showed more punitive responses when given the opportunity to give feedback to the "child" at the end of the "training" session (Bugental, Lewis, et al., 1999).

In summary, adults with low perceived power when experimentally paired with or primed for interaction with a difficult child show a variety of responses that reflect a defensive response pattern in terms of their communication and their behavior. This set of investigations gave a strong focus to the role of cognitions within transactional relationships. In addition, we found that the effects observed in caregiver interactions could only be accounted for by the interaction between adults' beliefs and children's characteristics. Indeed, no significant main effects were found for either caregivers' beliefs or children's characteristics.

Biological Mediators of the Relationship Between Parental Cognitions and Behavior

As support emerged for the cognitive factors that foster maladaptive and hostile responses, other questions arose. In particular, it became important to determine the mediating variables. Did harsh parenting follow simply from biased information processing, or did those biases lead to some kind of physiological response that more directly accounted for parental reactions? Parental responses typically appeared to be defensive in nature—as if these individuals were protecting themselves against a threat. Was it possible that physiological threat response systems were exceptionally easily activated among individuals who themselves might be thought of as *threat sensitive?* And was their targeting of children whose response characteristics were ambiguous a significant factor? Relevant to this question, it should be noted that some biological response systems are activated in stressful situations in which there might be a threat, and preparatory processes are engaged for this eventuality (as described by McEwen, 2002). Possibly parents who are threat sensitive are more likely to be continuously vigilant for potential threats within caregiving relationships.

Over time, increasing advantages have been found for the inclusion of biological factors within social relationships. Consideration of transactional processes in families has broadened to include these variables. For example, developmental neuroscience has provided guidance with respect to some of the physiological processes that serve to mediate transactions in parenting relationships (e.g., Bugental, Olster, & Martorell, 2002). Cognitions influence behaviors not only as a function of selective information processing but also through mediating processes that occur at the level of the central nervous system (e.g., threat-related hormonal changes).

As anticipated, threat-oriented adults, when paired with a difficult child, showed high activation of physiological stress response systems. For example, they showed increases in cortisol levels (Lin et al., 2002) and increases in heart rate (Bugental et al., 1993; Bugental, Lewis, et al., 1999; Lin et al., 2002). In addition (and unlike other adults), they failed to habituate (in terms of heart rate) when repeatedly exposed to the same stimulus children (Bugental, Olster, & Martorell, 2002). In other research, they were also found to show increases in electrodermal activity (Bugental et al., 1993; Bugental & Cortez, 1988) as well as decreases in (finger) skin temperature (Bugental & Cortez, 1988). This general pattern is consistent with the physiological reactions expected to be shown in response to perceived threat. The observed increases in heart rate were also found to serve as a partial mediator of the punitive responses shown to (computer-simulated) children (Bugental, Lewis, et al., 1999; Bugental, Olster, & Martorell, 2002).

In a similar fashion, mothers with their own infants have been found to show increasing physiological activation as a function of their biased perceptions and the difficult nature of their child's temperament (Martorell & Bugental, 2006). Mothers with low perceived power were more likely to show increased cortisol levels in response to a stressful interaction when their children also revealed a difficult temperament pattern. This response pattern, in turn, was found to mediate these mothers' greater use of harsh parenting practices.

Child Reactions to the Response Style of Threat-Sensitive Caregivers

Bringing the process back full circle (within a transactional framework), measures were taken of the reactions of children to the response style of adults with high or low perceived power (Bugental, Lyon, Lin, McGrath, & Bimbela, 1999). In two different studies, children believed they were receiving training on a task from an unrelated adult (whom they saw on videotape). The measures of primary interest were their orienting responses while watching the adult and their subsequent performance on a cognitively demanding task (mental arithmetic). In the first study, the tapes were those made of previous research participants when they believed they were training a child. In the second study, the tapes involved an actress who enacted the nonverbal responses found to be typical of women with low perceived power (when paired with difficult children) versus the nonverbal responses more typically shown by other women. In both cases, the communication style of the low-power teacher was ambiguous (e.g., non-Duchenne smiles and uncertain voices combined with relatively assertive verbal content). Children responded in significantly different ways to the two communication patterns. That is, they were more likely to orient to the nonambiguous teacher (e.g., showing reduced heart rate) than to the ambiguous teacher. Following their interaction with an ambiguous teacher, children showed greater performance deficits on a cognitively demanding task. These findings suggest that the response pattern more common for adults with low perceived power ultimately feeds back to decrease the attention and performance levels of such children, thus increasing the extent to which they could be seen as difficult. In a second study (Bugental & Happaney, 2000), we observed that children exposed to the derogating speech of a threat-sensitive adult (either a parent or nonparent) were more likely to withdraw from that interaction.

It appears that children are more likely to withdraw or disengage from the response style of a threat-sensitive adult. Thus, the response style of threat-sensitive adults tends to elicit responses from children that ultimately serve to confirm the expectations of those adults.

LONGITUDINAL RESEARCH IN FAMILIES

Although the experimental program described allowed the possibility of drawing causal inferences—and thus providing support for the proposed model in terms of internal validity—it lacked external validity. Still needed was evidence that equivalent processes are actually found in families that are at risk of child mistreatment or maltreatment. What was needed was a longitudinal study that tracked the transactional processes that emerged during infancy. A full test of the model required the assessment of parental attributions before the child was born, thus allowing a test of the extent to which initial adult attributions serve to differentially moderate their reactions to children who might be interpreted as presenting a caregiving threat or challenge at birth (vs. children who posed no such threat).

A short-term longitudinal study was conducted (Bugental & Happaney, 2004) of families identified as at risk of maltreatment (because of the background of these parents). Of the children within these families, 31% were born at some level of medical risk; that is, they were born prematurely or had relatively low (5-minute) Apgar scores. Prior to the birth of children, mothers completed the Parent Attribution Test. The primary outcome measure was the mother's use of harsh parenting during the 1st year of life. At the end of the year, mothers completed the Conflict Tactics Scale (Straus, 1979), a commonly used measure of child maltreatment or mistreatment. In addition, mothers completed the Beck Depression Inventory (Beck, Steer, & Garbin, 1988).

The combination of child medical risk at birth (primarily preterm status) and the mother's prebirth attributions served as a significant predictor of (a) maternal depression, (b) maternal use of harsh practices (primarily spanking or slapping), and (c) maternal neglect in maintaining child safety. No significant differences or trends were found as a result of maternal parity. Thus, it appears that maternal attributions were predictors rather than reflections of their history with a particular child. Maternal depression, in turn, partially mediated the observed relation between predictor variables and harsh parenting.

Further research was conducted to determine the reactions of infants to the responses found to be characteristic of threat-sensitive mothers interacting with infants who experienced birth complications. As toddlers, children were placed in a fear-inducing situation (the Strange Situation), and their cortisol levels were measured (both at baseline and in reaction to separation from their mothers). The highest level of cortisol reactivity was shown by infants whose mothers had demonstrated harsh tactics during their 1st year of life (Bugental, Martorell, & Barraza, 2003). In addition, children manifested variations in their basal levels of cortisol as a function of maternal depression. Basal levels were twice as high among children of mothers who manifested depressive symptoms.

Ultimately, the hormonal responses shown by children have important implications for their later outcomes. Overactivation of the hypothalamic–pituitary–adrenal axis (either as a function of excess basal levels of cortisol or high cortisol reactivity) predicts a number of later negative outcomes, for example, deficits in memory, affective and behavioral problems, immune system dampening and health problems, interference with growth and reproduction, and high reactivity and low habituation to stress or novelty (Bremner & Narayan, 1998; Sapolsky, 1996). Any one of these outcomes also serves to heighten the challenge that such children pose for their parents.

MALTREATMENT PREVENTION IN HIGH-RISK FAMILIES

One of the strongest possible tests of a transactional model is the extent to which it can be useful in designing a program that effectively prevents the psychopathology-amplifying effects that follow when mothers with threat-focused explanatory biases are paired with children who pose a potential threat or challenge to their caregiving ability and maintenance of authority. To provide such a test, an intervention was designed that was partially modeled after existing home visitation programs extended at birth to at-risk families. However, the central features of the program were specifically designed to allow cognitive reframing of the challenges presented by difficult infants (Bugental, Ellerson, et al., 2002).

Rather than creating an entirely new program, new features were added to the basic healthy start model originally introduced in Hawaii (Breakey & Pratt, 1991; Duggan et al., 1999). Families were referred by local hospitals and health care agencies on the basis of traditional risk factors, for example, poverty, stress, absence of social support, low education, and maternal history of having been abused herself. Families accepted to the program scored as being at moderate risk on the Family Stress Checklist (Murphy, Orkow, & Nikola, 1985). Families were randomly assigned to one of three different conditions:

1. Group C: This was a control condition in which parents were given information concerning existing services available to them in the community (but no provision of direct services was offered).
2. Group B: The second condition essentially used the Healthy Start model and focused on traditional parent education, provision of social support, and assisting the formation of linkages with others in the community.
3. Group A: The third condition included the features offered in Groups B and C but in a way that focused on cognitive reframing. That is, instead of offering solutions, information or guidance, mothers were assisted in learning how to become effective information seekers and problem solvers.

Because the third intervention condition was novel, it may be useful to provide more details on the method used. At each visit, mothers were asked to describe how things had been going with the infant since the last visit, including any problems or concerns. When the mother identified a problem, she was then asked what she thought was the cause. If the mother identified a cause that involved blame (of self or others), she was encouraged to think of other possible causes. When a cause was considered that was not blame oriented, she was then asked to think about possible resolutions she could try. In the consideration of possible causes and resolutions, mothers were always encouraged to think about how one knows what an infant is experiencing in view of the fact that he or she cannot tell one. This procedure served to facilitate mothers' own problem solving, and no direct advice or information was offered. For example, mothers might have been encouraged to locate the books and videotapes the state (California) had provided them; the home visitor and the mother then explored together what was known about the problem identified by the mother along with some of the possible solutions offered.

As a specific example, one mother thought her infant was "talking back" to her at 3 months of age, reflecting the commonly observed process in which infants babble in response to hearing their mother talk. The mother, however, had interpreted the infant's behavior as an early warning of later problems. By exploring the possible explanations of this "problem," she was able to reconstrue the meaning of the event. She was also able to observe that the infant, in doing this, was also showing happy affect.

Families in Groups A and B received the same number of home visits across the 1st year of the child's life. Significantly greater effects were found in the A than in the B or C conditions. The lowest levels of abuse (4%) were found in the A condition (in comparison with 23% in the B condition and 26% in the C condition). Similar advantages were found in reductions in maternal use of nonabusive but harsh tactics (e.g., spanking) and in the reports of child injuries (factors that may either reflect neglect or that may serve as a subtle marker of unreported abuse).

In addition, lower levels of maternal depression (at the end of a year) were found in the A condition than in the B or C conditions. Reductions in maternal depression were found to serve as a mediator of the observed reductions in harsh parenting tactics found in the cognitive reframing condition.

A replication of the cognitive reframing program was conducted that included an assessment of maternal practices (Bugental & Schwartz, in press). A comparison was again offered between the A condition and the B condition. Referral of families to the program, rather than being based on the parent's history, was now based on the child's medical history. Families were included in which infants had experienced some type of medical problems at birth (conditions that predict elevated risk for abuse; Brown, Cohen, Johnson, & Salzinger, 1999; Sherrod et al., 1984).

Previous findings were replicated in terms of the advantage of the cognitive reframing condition for reducing harsh parenting practices. Mothers in the A condition were less likely to use harsh parenting practices than were mothers in the B condition. Consistent with this pattern, fewer injuries were reported during the 1st year of life for infants in the A condition than the B condition.

DIFFERENTIAL REACTIVITY OF CHILDREN TO SOCIAL CONTEXT

At some point within this program of research, it became apparent that not all children were equally reactive to their social context. As suggested by Boyce and Ellis (2005), children differ in their reactivity to social context. For example, the stressful nature of a child's early experiences may foster elevated levels of reactivity. Belsky and his colleagues (Belsky, 2005; Belsky, Hsiah, & Crnic, 1998) as well as Boyce and Ellis (2005) have also suggested that children differ in their sensitivity to their parenting history on the basis of their temperament patterns. Differential reactivity of the young to their early experiences is not limited to humans, however; variations in the reactivity of the young to maternal care have also been observed among nonhumans (e.g., Liu, Diorio, Day, Francis, & Meaney, 2000).

In two studies within our own program of research, children with more severe problems at birth were found to manifest both the most and the least adaptive hormonal responses on the basis of their social environment. In a retrospective study (Bugental et al., in press), young adults who had experienced early medical problems were found to respond differentially to later stressors on the basis of their ongoing level of social support. Young adults who experienced early medical problems but currently had a high level of social support were more likely to show higher levels of habituation (reduced cortisol reactivity to a stressor when it was repeated) than did others, including those who lacked any history of medical problems and who currently had high social support.

In prospective research (Bugental, Beaulieu, & Schwartz, 2008), infants who were born prematurely were shown to be more reactive to maternal depression than were full-term infants. That is, the highest levels of basal cortisol were shown by preterm children in response to maternal depression. At the same time, the very lowest levels of basal cortisol were shown by preterm children in response to nondepressed mothers.

INTEGRATION

The model we tested with respect to parenting processes in high-risk families began with an emphasis on the role of parental cognitions as moderators of the differential responses shown to difficult children. Our early research demonstrated that the transactional processes involved were reciprocal in nature. The kinds of responses shown by threat-oriented parents to difficult children fed back to increase the likelihood that such children would continue (or escalate) in the potential threat they offered to vulnerable parents. In addition, pathology amplification or attenuation in families depends not only on the differential sensitivity of parents to caregiving threat but also on the differential sensitivity of children to their social context. The processes observed are consistently reflected in interactions rather than main effects. In pathology-amplifying situations, the combination of a threat-sensitive parent and a context-sensitive child creates a negative transactional system. In contrast, a pathology-attenuating situation is fostered with the combination of (a) a parent who sees anomalous caregiving situations as challenges rather than threats and (b) a context-sensitive child. Context sensitivity in children leads to heightened reactivity to positive contexts (with resultant pathology attenuation) as well as heightened reactivity to negative contexts (with resultant pathology amplification).

Extending this model (as a biocognitive transactional model), we went on to demonstrate relevant mediating processes at a physiological level. That is, when parents were confronted with ambiguous threat (as a result of the characteristics of the child and their own explanatory biases), they were more likely to manifest threat-consistent physiological responses (e.g., increases in heart rate and cortisol production). These changes in turn were found to mediate the harsh responses shown by such parents. Parental harshness (in particular, when combined with depression) fostered increases in the stress responses of infants—in particular among children at greatest risk (i.e., premature children). Thus, the inclusion of measures and concepts from developmental neuroscience may aid in explaining transactional processes in families.

As a future direction, we are exploring the transactional processes that occur in the reciprocal effects of biology and life experience. These two approaches—previously seen as opposed—are increasingly being understood in an integrated fashion.

One biological view (an evolutionary approach) focuses on the extent to which the young (and their parents) are designed for adaptive responses to problems that have recurred across a species' evolutionary history. For example, environments vary over time in the resources available for care of the young. Parental investment (among both humans and nonhumans) in "costly" infants (e.g., those born with cues to health risk) is relatively low when parents lack resources (material or social). On the other hand, parental investment in costly infants may be exceptionally high when those parents have high access to resources and thus can afford the added investment in the costly offspring without putting siblings in peril. This position was suggested by Mann (1992) and was subsequently amplified and empirically demonstrated by Bugental and Beaulieu (Beaulieu & Bugental, 2006, 2008; Bugental & Beaulieu, 2003).

The second view focuses on the extent there is change in the young (at the level of brain and behavior, as mediated by differential gene expression) in response to the nature of their early environment (Liu et al., 2000). If the young experience high stress—without maternal responses that reduce that stress—they show gene expression that fosters fearfulness. However, if the young experience maternal responses that reduce their stress (for rat pups, this involves licking and grooming, and arched-back nursing), they show more adaptive gene expression (and behavior) at later ages. As a result of this second transactional system, the young accommodate within their lifetime to the actualities of the environment into which they are born.

The two approaches are now understood as working well together as an adaptive transactional process. As biological models are receiving greater attention within the study of parenting relationships, increasing recognition is being given to the fact that the brain is both *experience dependent* (as demonstrated in work on differential gene expression) and *experience expectant* (as promoted by an evolutionary framework)—an integrative framework suggested by Bruer & Greenough (2001). Consideration of these fully compatible and interwoven processes extends the notions of transactions in parental relationships in a new and highly productive direction. The combined transactional processes are not only important from a theoretical standpoint, they also provide guidance in the design of remedial programs that foster pathology attenuation within at-risk family systems.

In conclusion, the transactional framework originally proposed by Sameroff has proven to be flexible in allowing the incorporation of ideas drawn from a variety of theoretical perspectives as well as the full range of variables thought to operate within families. It has also provided a valuable way of understanding the processes that operate in pathology-attenuating as well as pathology-amplifying family systems.

REFERENCES

Beaulieu, D. A., & Bugental, D. B. (2006). An evolutionary approach to socialization. In J. Grusec & P. Hastings (Eds.). *Handbook of socialization: Theory and research*. New York: Guilford Press.

Beaulieu, D. A., & Bugental, D. B. (2008). Contingent parental investment: An evolutionary framework for understanding early interaction between mothers and children. *Evolution and Human Behavior, 29*, 249–255.

Beck, A., Steer, R., & Garbin, M. (1988). Psychometric properties of the Beck Depression Inventory: Twenty-five years of evaluation. *Clinical Psychology Review, 8*, 77–100.

Bell, R. Q. (1968). A reinterpretation of the direction of effects in studies of socialization. *Psychological Review, 75*, 81–95.

Belsky, J. (1984). The determinants of parenting: A process model. *Child Development, 55*, 83–96.

Belsky, J. (2005). Differential susceptibility in rearing influence: An evolutionary hypothesis and some evidence. In B. Ellis & D. Bjorklund (Eds.), *Origins of the social mind: Evolutionary psychology and child development* (pp. 139–163). New York: Guilford Press.

Belsky, J., Hsiah, K-H., & Crnic, K. (1998). Mothering, fathering, and infant negativity as antecedents of boys' externalizing problems and inhibition at age 3 years: Differential susceptibility to rearing experience? *Development and Psychopathology, 10*, 301–319.

Boyce, W. T., & Ellis, B. J. (2005). Biological sensitivity to context: 1. An evolutionary-developmental theory of the origins and functions of stress reactivity. *Development and Psychopathology. 17*, 271–301

Breakey, B., & Pratt, B. (1991). Healthy growth for Hawaii's Healthy Start: Toward a systematic statewide approach to the prevention of child abuse and neglect. *Zero to Three, 11*, 16–22.

Bremner, J. D., & Narayan, M. (1998). The effects of stress on memory and the hippocampus throughout the life cycle: Implications for childhood development and aging. *Development and Psychopathology, 10*, 871–885.

Brown, J., Cohen, P., Johnson, J. G., & Salzinger, S. (1999). A longitudinal analysis of risk factors for child maltreatment: Findings of a 17-year prospective study of officially recorded and self-reported child abuse and neglect. *Child Abuse and Neglect, 22*, 1065–107.

Bruer, J. T., & Greenough, W. T. (2001). The subtle science of how experience affects the brain. In D. B. Bailey, J. T. Bruer, F. J. Symons, & J. W. Lichtman (Eds.), *Critical thinking about critical periods* (pp. 209–232). Baltimore: Brookes Publishing.

Bugental, D. B. (1992). Affective and cognitive processes within threat-oriented family systems. In I. E. Sigel, A. McGillicuddy-DeLisi, & J. Goodnow (Eds.), *Parental belief systems: The psychological consequences for children* (pp. 219–248). New York: Erlbaum.

Bugental, D. B. (2003). *Thriving in the face of childhood adversity*. New York: Psychology Press.

Bugental, D. B., & Beaulieu, D. A. (2003). Maltreatment risk among disabled children: A bio-social-cognitive approach. In R. Kail (Ed.), *Advances in child development and behavior* (Vol. 31, pp. 129–164). New York: Academic Press.

Bugental, D. B., Beaulieu, D., O'Brien, E., Schwartz, A., Cayan, L., Fowler, E., & Ellerson, P. (in press). Reactivity to stress: When does a history of medical adversity foster resilience versus vulnerability? *Journal of Applied Social Psychology*.

Bugental, D. B., Beaulieu, D. A., & Schwartz, A. (2008). Hormonal sensitivity of preterm versus full-term infants to the effects of maternal depression. *Infant Behavior and Development, 31*,51–61.

Bugental, D. B., Blue, J., Cortez, V., Fleck, K., Kopeikin, H., Lewis, J., & Lyon, J. (1993). Social cognitions as organizers of autonomic and affective responses to social challenge. *Journal of Personality and Social Psychology, 64*, 94–103.

Bugental, D. B., Blue, J., & Cruzcosa, M. (1989). Perceived control over caregiving outcomes: Implications for child abuse. *Developmental Psychology, 25*, 532–539.

Bugental, D. B., Blue, J., & Lewis, J. (1990). Caregiver beliefs and dysphoric affect directed to difficult children. *Developmental Psychology, 26*, 631–638.

Bugental, D. B., Brown, M., & Reiss, C. (1996). Cognitive representations of power in caregiving relationships: Biasing effects on interpersonal interaction and information processing. *Journal of Family Psychology, 10*, 397–407.

Bugental, D. B., Caporael, L., & Shennum, W. A. (1981). Experimentally produced child uncontrollability: Effects on the communication of adults with different control perceptions. *Child Development, 51*, 520–528.

Bugental, D. B., & Cortez, V. L. (1988). Physiological reactivity to responsive and unresponsive children as moderated by parental attributions. *Child Development, 59*, 686–693.

Bugental, D. B., Ellerson, P. C., Lin, E. K., Rainey, B., Kokotovic, A., & O'Hara, N. (2002). A cognitive approach to child abuse prevention. *Journal of Family Psychology, 16*, 243–258.

Bugental, D. B., & Happaney, K., (2000). Parent–child interaction as a power contest. *Journal of Applied Developmental Psychology, 21*, 267–282.

Bugental, D. B., & Happaney, K. (2004). Predicting infant maltreatment in low-income families: The interactive effects of maternal attributions and child status at birth. *Developmental Psychology, 40*, 234–243.

Bugental, D. B., Kopeikin, H., & Lazowski, L. (1991). Children's responses to authentic versus polite smiles. In K. J. Rotenberg (Ed.), *Children's interpersonal trust* (pp. 58–79). New York: Springer-Verlag.

Bugental, D. B., & Lewis, J. (1998). Interpersonal power repair in response to threats to control from dependent others. In M. Kofta, G. Weary, & G. Sedek (Eds.), *Personal control in action: Cognitive and motivational mechanisms* (pp. 341–362). New York: Plenum Press.

Bugental, D. B., & Lewis, J. C. (1999). The paradoxical misuse of power by those who see themselves as powerless: How does it happen? *Journal of Social Issues, 55*, 51–64.

Bugental, D. B., Lewis, J. C., Lin, E., Lyon, J., & Kopeikin, H. (1999). In charge but not in control: The management of authority-based relationships by those with low perceived power. *Developmental Psychology, 35*, 1367–1378.

Bugental, D. B., Lyon, J. E., Cortez, V., & Krantz, J. (1997). Who's the boss? Accessibility of dominance ideation among individuals with low perceptions of interpersonal power. *Journal of Personality and Social Psychology, 72*, 1297–1309.

Bugental, D. B., Lyon, J. E., Lin, E., McGrath, E. G., & Bimbela, A. (1999). Children "tune out" in response to the ambiguous communication style of powerless adults. *Child Development, 70*, 214–230.

Bugental, D. B., Martorell, G. A., & Barraza, V. (2003). The hormonal costs of subtle forms of infant maltreatment. *Hormones and Behavior, 43*, 237–244.

Bugental, D. B., Olster, D. H., & Martorell, G. A. (2002). A developmental neuroscience perspective on the dynamics of parenting. In L. Kuczynski (Ed.), *Handbook of dynamics in parent–child relationships* (pp. 25–48). Thousand Oaks, CA: Sage.

Bugental, D. B., & Schwartz, A. (in press). A cognitive approach to child maltreatment prevention. *Developmental Psychology*.

Bugental, D. B., & Shennum, W. A. (1984). "Difficult" children as elicitors and targets of adult communication patterns: An attributional–behavioral transactional analysis. *Monographs of the Society for Research in Child Development, 49*(1, Serial No. 205).

de Lissovoy, D. (1979). Toward the "abuse-provoking child." *Child Abuse and Neglect, 3*, 341–350.

Duggan, A., McFarlane, E. C., Windhan, A. M., Rohde, C. A., Salkever, D. S., Fuddy, L., et al. (1999) Evaluation of Hawaii's Healthy Start program. *Future of Children, 9*, 66–90.

Elmer, E., & Gregg, G. S. (1967). Developmental characteristics of abused children. *Pediatrics, 40,* 596–602.

Francis, D., Diorio, J., Liu, D., & Meaney, J. (1999). Nongenomic transmission across generations of maternal behavior and stress responses in the rat. *Science, 286,* 1155–1158.

Lewis, J. C., Bugental, D. B., & Fleck, K. (1991). Beliefs as moderators of reactions to computer-simulated responsive and unresponsive children, *Social Cognition, 9,* 277–293.

Lin, E. K., Bugental, D. B., Turek, V., Martorell, G. A., & Olster, D. H. (2002). Children's vocal properties as mobilizers of stress-related physiological responses in adults. *Personality and Social Psychology Bulletin, 28,* 346–357.

Liu, D., Diorio, J., Day, J. C., Francis, D. D., & Meaney, M. J. (2000). Maternal care, hippocampal synaptogenesis and cognitive development in rats. *Nature Neuroscience, 3,* 799–806.

Maestripieri, D. (2005). Early experience affects the intergenerational transmission of infant abuse in rhesus monkeys. *Proceedings of the National Academy of Sciences, 102,* 9726–9729.

Mann, J. (1992). Nurturance or negligence: Maternal psychology and behavioral preference among preterm twins. In J. H. Barkow, L. Cosmides, & J. Tooby, (Eds.), *The adapted mind: Evolutionary psychology and the generation of culture* (pp. 367–390). Oxford, England: Oxford University Press.

Martorell, G. A., & Bugental, D. B. (2006), Maternal variations in stress reactivity: Implications for harsh parenting practices with very young children. *Journal of Family Psychology, 20,* 641–647.

McEwen, B. S. (2002). *The end of stress as we know it.* Washington, DC: Joseph Henry Press.

Murphy, S., Orkow, B., & Nicola, R. M. (1985). Prenatal prediction of child abuse and neglect: A prospective study. *Child Abuse and Neglect, 9,* 225–235.

Sameroff, A. J. (1975). Early influences on development: Fact or fancy? *Merrill-Palmer Quarterly, 21,* 267–294.

Sameroff, A. J., & Chandler, M. J. (1975). Reproductive risk and the continuum of caretaking casualty. In F. D. Horowitz, M. Hetherington, S. Scarr-Salapatek, & G. Siegal (Eds.), *Review of child development research* (Vol. 4, pp. 187–244). Chicago: The University of Chicago Press.

Sameroff, A. J., & Emde, R. N. (Eds.). (1989). *Relationship disturbances in early childhood: A developmental approach.* New York: Basic Books.

Sameroff, A. J., & MacKenzie, M. J. (2003). Research strategies for capturing transactional models of development. The limits of the possible. *Development and Psychopathology, 15,* 613–640.

Sapolsky, R. M. (1996). Stress, glucocorticoids, and damage to the NS: The current state of confusion. *Stress, 1,* 1–19.

Sherrod, K. B., O'Connor, S., Vietze, P. M., & Altemeier, W. A., III. (1984). Child health and maltreatment. *Child Development, 55,* 1174–1183.

Spinetta, J. J., & Rigler, D (1972). The child-abusing parent: A psychological review. *Psychological Bulletin, 77*, 296–304.

Steele, B. F., & Pollock, D. D. (1968). A psychiatric study of parents who abuse infants and small children. In R. E. Helfer & C. H. Kempe (Eds.), *The battered child*. Chicago: The University of Chicago Press.

Straus, M. A. (1979). Measuring intrafamily conflict and violence: The Conflict Tactics (CT) Scales. *Journal of Marriage and the Family, 41*, 75–88.

Thomas, A., Chess, S., & Birch, H. G. (1968). *Temperament and behavior disorders in children*. New York: New York University Press.

Wong, P. T., & Weiner, B. (1981). When people ask "why" questions, and the heuristics of attributions. *Journal of Personality and Social Psychology, 40*, 650–663.

7

SOCIAL INFORMATION PROCESSING AND AGGRESSIVE BEHAVIOR: A TRANSACTIONAL PERSPECTIVE

REID GRIFFITH FONTAINE AND KENNETH A. DODGE

Transactional theory articulates an ongoing reciprocal exchange between the self and one's experiences with his or her social world. Although the self in this theory has been typically operationalized as behavior, it clearly must include cognitive operations. Social information processing (SIP) theory posits that behavior toward others in childhood and adolescence develops as a function of the outputs of complex sets of online social–cognitive operations. In transactional turn, behavior leads to social consequences that inform future SIP. In this way, SIP and interactions with others may influence each other across development. One set of SIP operations that appears to be particularly important to adolescent development is called *response evaluation and decision* (RED). RED processing is an advanced stage of SIP in which an individual evaluates alternative responses across multiple domains to decide how to respond to cues during social interaction. In this chapter, we present a conceptual model of the transactional development of SIP with special

Support for this research was provided by the Child Development Project, which has been funded by Grants MH42498, MH56961, MH57024, and MH57095 from the National Institute of Mental Health; HD30572 from the National Institute of Child Health and Human Development; DA016903 from the National Institute on Drug Abuse; and 5P20 DA017589 from the Duke Transdisciplinary Prevention Research Center.

application to understanding antisocial behavior in youth. We highlight the role of RED in this developmental exchange during adolescence and identify future directions for research on the development of social cognition and antisocial behavior reciprocations.

TRANSACTIONS BETWEEN THE SELF AND THE SOCIAL WORLD

Transactional theory posits that an individual's development is the product of ongoing bidirectional influences on the individual and his or her environment (Sameroff, 2000; Sameroff & Chandler, 1975). This perspective has been used to explain patterns of both adaptive and maladaptive development (Sameroff & MacKenzie, 2003). For example, the child who is abused may, as a result, develop a behavioral pattern by which he or she acts out and defies authority more regularly. In transactional turn, the child may be the recipient of harsher, more frequent beatings. A similar example involves peer relationships. The child who is peer rejected may misbehave to get attention or aggress against his or her peers out of anger and frustration. As a result, his or her peers may be even less accepting, and the child may become identified as a victim target.

Over the last 3 decades, scientists have advanced transactional models of maladaptive development in childhood and adolescence (e.g., Cicchetti, Toth, & Maughan, 2000; Crick & Dodge, 1994; Sameroff & Chandler, 1975). Developmental and phenomenological foci have ranged from the reciprocal influence of parent stress and responsiveness and infant temperament (Shaw & Winslow, 1997) to transactions of parent depressive symptoms and adolescent behavior problems and self-esteem (Killeen & Forehand, 1998). In particular, approaches that recognize person–environment transactions in the study of antisocial development have been stressed (Dodge & Pettit, 2003; Granic & Patterson, 2006). The theoretical richness of current transactional perspectives is undeniable, though rigorous empirical investigations of these models remain relatively few. Because antisocial conduct creates significant problems for those around the actor, this style of behavior continues to be a significant social concern. It remains essential to investigate how antisocial behavior, through dynamic exchange with other individual systems (e.g., cognition, emotion) and environment, may account for alternative developmental trajectories by which a multifinal range of adaptive and maladaptive outcomes across youths may be explained.

One particular form of transaction is that between the self's social cognitions and his or her behavioral interactions with the world. A transactional perspective of social cognition and behavior maintains that social–cognitive operations that occur in real time (including the interpretation of social cues and evaluative judgment of alternative behavioral options) cause social behaviors, and that, in turn, social behaviors exact experiences and consequences in the world that inform future social–cognitive processing. The critical idea is that environmental factors do not affect a person's future behavior directly but rather are mediated by ongoing cognitive operations that organize incoming information so that it may be used to guide subsequent behavior. Current transactional models of antisocial development have acknowledged the role of social cognition in varying ways and to varying degrees, from a modest acknowledgement of the role of cognitive appraisals (Granic & Patterson, 2006) to an elaborate formulation of the interplay between multiple series of online processes and a dynamic network of mental structures stored in memory (Crick & Dodge, 1994).

The latter of these perspectives, SIP (Crick & Dodge, 1994), asserts that youth antisocial behavior is the product of a complex set of social–cognitive processes that are active as the individual receives and reads information from his or her environment and prepares to respond to social cues. Five sets of processes have been posited. First, the youth encodes and organizes features of the social stimulus (*encoding of cues*). Second, attributions of affect, intent, and causality are made (*interpretation of cues*). Next, the youth identifies and prioritizes his interests in the given situation (*clarification of goals*). Fourth, one or more responses to the cue are generated, either by accessing them from memory or by constructing them anew (*response access or construction*). These responses are then evaluated across multiple domains such as the likely success of carrying out a response and what outcomes may result (*response decision*). Ultimately, a response is selected for behavioral enactment. Processes at each step of SIP theory have been empirically supported (see Crick & Dodge, 1994) and, collectively, are hypothesized to interrelate, relate to other individual systems (e.g., physiology), and transact with an individual's environment in a nonlinear, dynamic fashion (Crick & Dodge, 1994; Dodge & Pettit, 2003; Fontaine, 2006a).

SIP research has identified numerous individual-differences social–cognitive variables that distinguish antisocial from socially competent youths. Antisocial youths demonstrate biased patterns of processing at each SIP step that, collectively, account for greater variance in youth antisocial behavior than any subset of processing steps or factors. For example, as compared with his or her socially competent peers, the aggressive youth who is ambiguously provoked (perhaps by being bumped into on the playground) may be more likely to (a) encode and organize less social information and fewer social cues (Dodge & Newman, 1981); (b) attribute hostile intent to ambiguous provocateurs (e.g., Guerra & Slaby, 1989), (c) have goals that are potentially detrimental to social relationships (e.g., Slaby & Guerra, 1988); (d) access aggressive response options and fewer total alternative responses (e.g., Asarnow & Callan, 1985); and (e) evaluate aggressive responses favorably, both as a style of conduct and as a means to desired outcomes (see Fontaine & Dodge, 2006). In the case of an aggressive youth colliding with a peer in the school yard, the aggressive youth may use limited information to make the attribution that the peer was intentionally trying to hurt or embarrass him or her, clarify that he or she does not want to appear soft or weak to others, generate and favorably evaluate an aggressive response, and select it for retaliatory enactment.

Huesmann (1998) advanced a social–cognitive information-processing model that shares many features with the SIP model but is distinguished, in part, by its focus on script-based processing (cf. Fontaine & Dodge, 2006, in which we discussed strict impulsivity as the automatic processing of stored scripts). Scripts contain both procedural and declarative knowledge that may be used to guide expectations as to what will happen in one's environment and how to respond to these environmental happenings if and when they actualize. Huesmann asserted that these scripts are formed early in development through controlled cognitive processing in the context of ongoing social interaction. After scripts have developed, they are accessed, evaluated, selected, and enacted through automatic processing that requires significantly fewer mental resources. According to Huesmann, scripts are reevaluated and refined after they have been behaviorally enacted and an individual reflects on them in the context of their corresponding social consequences. The very nature of this proposition is transactional, and in this way, Huesmann's model is consistent with that of Crick and Dodge and other aggression researchers who have hypothesized that development is based on real-time transactions between cognitively mediated individual behaviors and environmental responses (e.g., Fontaine & Dodge, 2006; Granic & Patterson, 2006).

JUVENILE VIOLENCE

Recent decades have witnessed a persistent pattern of serious juvenile violence (e.g., Kalogerakis, 2003). The behavioral sciences have responded with numerous efforts in basic and intervention science to understand and reduce adolescent antisocial behavior. One scientific trend is the increase in attention paid to general cognitive functioning and decision making (e.g., Choudhury, Blakemore, & Charman, 2006; Fried, & Reppucci, 2001; Guerra & Slaby, 1990; Steinberg, 2005; Steinberg & Cauffman, 1996; Steinberg & Scott, 2003) as well as specific SIP mechanisms of aggression (e.g., Dodge, Lochman, Harnish, Bates, & Pettit, 1997; Fontaine, Burks, & Dodge, 2002) in community and violent offender samples of adolescents. Adolescence is marked by a number of unique qualities—developmental characteristics that suggest that individual SIP during this period may undergo notable change.

First, adolescence has been found to be a critical time for brain maturation and the development of executive function skills, including reasoning, evaluative judgment, and decision making (Choudhury et al., 2006; Cobb,

1992; Keating, 2004; Steinberg, 2005; Steinberg & Scott, 2003). Because adolescents are still developing biologically, psychologically, and socially, they may still be characterized by a degree of maturational elasticity, thus allowing for person–environment transactions that influence how they think, behave, and interact with respect to social phenomena to play a determining role in the formation of their adult outcomes. This reason alone would suggest that adolescence is a critical time to study social–cognitive and behavioral development through a transactional lens.

Second, during the very same period in which the adolescent is developing his or her identity, self-concept, and character, he or she is especially susceptible to peer influence (Hart & Yates, 1997; Harter, Bresnick, Bouchey, & Whitesell, 1997; Steinberg & Scott, 2003). In an attempt to balance the interests of developing one's independence as an individual and sustain important peer relationships, adolescence may mark a period during which greater self-reflection takes place as individuals challenge their online evaluations of social situations (per Crick & Dodge, 1994) and refine their social scripts (per Huesmann, 1998). Thus, there is reason to believe that adolescence is an important time to address the development of SIP at the level of the individual.

Third, adolescents face contexts that are novel in important ways (Cobb, 1992). As adolescents age, they are increasingly faced with contexts that adults confront on a more regular basis, such as decision-making opportunities and dilemmas regarding how to spend one's time (e.g., school, work, social), to what and to whom to be committed (e.g., vocational or professional responsibilities, romantic relationships), and new responsibilities (e.g., earning money, acquiring and maintaining adequate living accommodations). It seems unlikely that adolescents would already have scripts stored in memory that may be accessed and used to guide their responses to novel situations as they arise. As such, it is reasonable to hypothesize that because adolescents face novel, more adultlike contexts with new social challenges, controlled (as opposed to automatic) processing plays a crucial role in the mediation of the relation between environmental cues and individual behaviors. Online operations such as those posited by Crick and Dodge's (1994) SIP model may be particularly well equipped to account for the social–cognitive development that adolescents undergo to tackle these new challenges and obstacles as they present themselves.

EXPANDED MODEL OF SOCIAL INFORMATION PROCESSING

Recently, it has been hypothesized that this mediational course may be particularly influenced by the step of SIP that includes evaluative judgment and decision processes (Fontaine, 2007a; Fontaine & Burks, 1999; Fontaine, Yang, Dodge, Bates, & Pettit, 2008). Fontaine and Dodge (2006) elaborated on earlier formulations of the response decision step of SIP and posited multiple domains of a response decision-making model, RED, which posits several sets of online decision processes. First, the idea of a *primary threshold of acceptability* was introduced. This initial processing mechanism serves as a crude filter by which the first response that is generated may be impulsively selected for behavioral enactment, such that it essentially bypasses subsequent RED processing or, alternatively, can filter out clearly inadequate response options. A responding youth may quickly accept and enact the first behavior that comes to mind for a number of reasons. For one, Huesmann's SIP model suggests that individuals have well-rehearsed scripts that are "on tap" and may be prominently positioned to activate when specific sets of social cues are identified. In addition, various contextual factors may contribute to impulsive responding, including the perception of social urgency, pharmacological or illicit substance influences, affect and mood, fatigue, hunger, other physiological states, and numerous other extracognitive factors.

Alternatively, the initial threshold of response acceptability may serve to quickly reject response options that are immediately discerned to be unacceptable. Fontaine and Dodge (2006) recognized multiple types of inadequate responses. First, a response may be irrelevant in that it does not apply to the given situation. Individuals with thought disorders have been shown to access (and often enact) response options that do not match the social exchange in which the person is interacting (e.g., see Roesch-Ely, Spitzer, & Weisbrod, 2003). Second, the response may be infeasible as a result of unique features of the social situation at hand. Perhaps a response schema is triggered because of typical aspects of a given interaction, but there exist other features of the interaction that are unusual, or at least different, that render the response option impractical. Third, a response option may be intrusive in other ways such that it is extreme. For example, the child who is teased may have an intrusive thought to stab the provocateur with his pencil but immediately rejects the idea, not because it is irrelevant or infeasible but because it is profoundly disproportionate to the level of harm that is perceived to be intended and caused by the provocateur.

Means-focused response judgments are also identified in the RED model. *Response efficacy* is the process by which a responding youth considers how likely it is that he or she will be able to successfully carry out a behavior being considered (Bandura, 1977). This process acts as a type of behavioral (as opposed to outcome) expectancy. Aggressive youths have been found to be significantly more confident in their aggressive actions than their non-aggressive peers (Erdley & Asher, 1996). Also, *response valuation* is an appraisal process through which the responding individual determines how much he or she identifies with the behavior in question as a social actor and moral agent (Fontaine et al., 2002; also, see Crick & Ladd, 1990; Guerra, Huesmann, & Hanish, 1995). A critical hypothesis of RED is that the aggressive youth's valuation of the sociomoral nature of a behavior is distinct from the outcomes with which the behavior is associated. The idea is that the degree to which a match is perceived between a considered behavior and an individual's personal identity (called *sociomoral congruence;* see Fontaine, 2007b) increases the likelihood of that behavior being further considered and ultimately enacted.

Expectancy-value models, such as that of Feather (1982), were also drawn on in the formulation of the RED model. Fontaine and Dodge (2006) recognized two subprocesses—*outcome valuation* and *outcome expectancy*. As opposed to the means-focused processes of response efficacy and valuation, outcome expectancy and outcome valuation refer to the ends that are in some manner associated with specific behaviors. Outcome valuation is the process by which there is some positive or negative appraisal of a potential result of acting in a specified way. Outcome expectancy is the estimated likelihood that *x* behavior will lead to *y* outcome. As compared with their nonaggressive peers, aggressive youths have been found both to value outcomes of aggressive behaviors more positively and to be more likely to expect that their aggressive actions will lead to desired results (e.g., see Deluty, 1983; Perry, Perry, & Boldizar, 1990; Slaby & Guerra, 1988). A critical lesson from traditional expectancy-value approaches is that the degree to which a specified outcome is anticipated to occur as a result of enacting a certain behavior is meaningful only to the degree that the outcome is positively or negatively valued. If no value (whether positive or negative) is assigned to the outcome, then the likelihood of its realization is naturally irrelevant to the responding youth's decision making.

In addition to expectancies of positive versus negative outcomes, Fontaine and Dodge (2006; Fontaine et al., 2008; also, see Smithmyer, Hubbard, & Simons, 2000) have hypothesized that alternative outcome expectancies, including expectancies of instrumental, social, and emotional outcomes, contribute to response decisions. With respect to a specific response, a behavior may be associated with both positive and negative outcomes across these qualitatively different domains. For example, acting aggressively may be judged to be highly likely to lead to the acquisition of desired material items (positive instrumental outcome) as well as feelings of guilt because the youth realizes that this is not a morally acceptable way to go about getting his or her wants met (negative emotional outcome). In addition, a specific response may be associated with both positive and negative outcomes within a domain. For example, a youth may expect that behaving aggressively will lead to peer acceptance as well as being "grounded" by his parents, impeding him from hanging out with his peers altogether (positive and negative social outcomes, respectively). There is empirical evidence that consideration of these alternative expectancy-value assignments may be particularly critical in the processing of instrumental (or proactive) aggressive behavior (Dodge et al, 1997; Smithmyer et al., 2000), and these processes have been incorporated into recent models of instrumental antisocial behavior in children and adolescents (Fontaine, 2006b, 2007b).

The final processes posited in the RED model are *response comparison* and *response selection*. During response comparison, the "best" overall response is identified on the basis of the values it has been assigned through appraisals of its sociomoral features and the outcomes with which the response is associated. The response that emerges as having the strongest total value is selected for behavioral enactment. Feedback loops are conceptualized to be active throughout RED processing to acknowledge a potentially infinite number of nonlinear trajectories of processing a behavioral decision in response to a social stimulus. Recognition of nonlinear patterns is reflected not only within the individual but also between the individual and his or her social environment, consistent with Crick and Dodge's (1994) reformulation of SIP. Person–environment transactions are hypothesized to serve as the basis for enactive learning (see Huesmann, 1998; Huesmann & Moise, 1998) and are, in this way, critical to a transactional perspective of the development of individual SIP and behavior.

Studies of response decision processes and antisocial behavior suggest that this set of processes may be critical to accounting for variance in externalizing conduct problems in adolescence. Fontaine, Burks, and Dodge (1998) found support for the hypothesis that sociomoral judgments about aggressive responses (a set of behavioral evaluations that may be likened to response valuation) mediate the relation between adolescent hostile attributional style (or the tendency to attribute hostile intent to ambiguous provocateurs) and antisocial behavior. Findings from this study supported the hypothesized processing-behavior sequence by which, in response to a perceived provocation, deviant adolescents attribute hostile intent to the ambiguous stimulus, evaluate aggressive retaliation as socially and morally acceptable or justified, and respond by enacting retributive aggression toward the perceived provocateur. A subsequent study (Fontaine et al., 2002) provided support for the distinct roles of means- and outcome-based evaluative judgments in Grade 9 adolescents' processing of aggressogenic (or aggression-eliciting) social cues. Response valuation and outcome expectancy were empirically differentiated as domains of RED, and each was found to uniquely increase the prediction of externalizing behavior. Clearly, RED processing plays an important role in how adolescents are affected by and respond to their social worlds.

TRANSACTIONS BETWEEN SOCIAL INFORMATION PROCESSING AND ANTISOCIAL BEHAVIOR

Contrary to the notion that one's processing patterns are static traitlike characteristics that wield unidirectional impact on behavior across the life course, the transactional perspective asserts that processing patterns evolve through reciprocal interactions between the self and one's social experiences. Processing patterns guide behavior and social experiences, but social experiences (including one's behavior and its immediate impact on others) also have an impact on how one processes future stimuli. Processing patterns, in turn, shape the selection of future social encounters and the child's behavior in that encounter. This ongoing transaction is depicted in Figure 7.1.

A model of ongoing reciprocal influences between the self and the social world does not imply that social cognitions change with every social encounter and have little stability across time. Rather, Dodge and Price (1994) found that SIP patterns do exhibit remarkable stability across years of elementary school. More interesting is that this stability emerges—that is, SIP patterns are not initially stable but become stable over time. What accounts for the emergence of stability?

Although social transactions alter future processing patterns, transactions may also account for the emergence of stability in processing patterns. Social encounters are not random but are partially selected by the self. A child who judges that aggressive behavior will yield desired instrumental outcomes and that prosocial behavior will yield ridicule by peers will be likely to be attracted to situations in which aggressive behaviors do, indeed, yield desired outcomes. That child may avoid situations that call for and reinforce prosocial action. The child's aggressive behavior (and success in achieving desired outcomes) in those situations reinforces the social–cognitive pattern that led to the aggressive behavior.

A similar pattern of selection in transactions may operate from the perspective of the peer group. Peers who encounter a child who behaves aggressively may begin to exclude that child from friendly social opportunities. Their encounters with the aggressive child may become restricted to exchanges in which prosocial behavior is not afforded and aggressive behavior becomes expected.

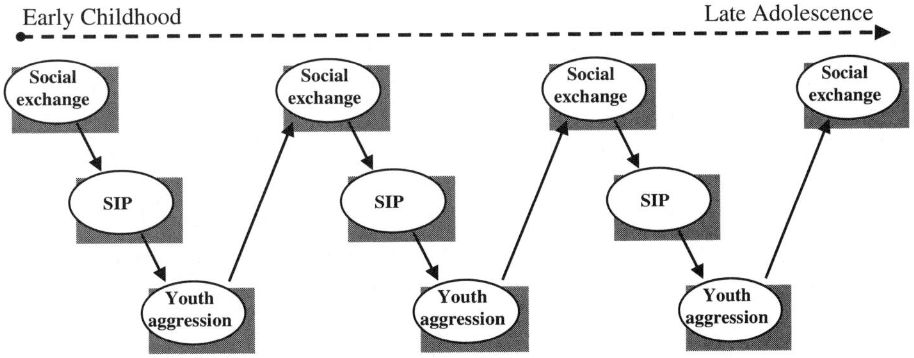

Figure 7.1. The role of social information processing (SIP) in the transactional development of youth aggressive behavior.

Over time, the child encounters fewer and fewer situations that could falsify the child's processing judgments. As time passes, the cycle of exchange between a child and his or her social environment (which is initially fairly undetermined) narrows until it becomes a vortex of hostility. Granic and Patterson (2006) described such exchanges between a child and his or her parent as a *coercive cycle;* similar exchanges must occur between the child and peers.

Although the transactional nature of social exchanges is likely to narrow the range of social encounters over time, when unexpected and cognitively discordant encounters do occur, they can have powerful influence in changing a child's behavior. Rabiner and Coie (1989) provided an example in which they interrupted the usual course of social exchange by experimental manipulation. They brought socially rejected aggressive youths into the laboratory and altered their usual processing patterns by telling them (or not, by random assignment) that a peer whom they were about to encounter would actually like them if they tried to interact nicely with the peer. Those rejected children whose processing patterns had been manipulated to believe that their behavior would lead to positive outcomes did, indeed, later fulfill that prophecy. Naive peers who interacted with these rejected children did like them more than they liked rejected children who had not been given this processing induction. Thus, the self-reinforcing pattern of increasing hostile exchanges can be interrupted by a sudden change in processing by the child.

These concepts have led to the design of clinical interventions that are intended to help aggressive children process the social world in a more benign manner. Hudley and Graham (1993) developed a 12-session group intervention that focused the attention of aggressive African American boys on the way in which they interpreted and evaluated social interactions, trying to move these boys toward less hostile and more benign interpretations and evaluations of the social world. Compared with a group of aggressive boys who had been randomly assigned not to receive the intervention, those boys who had received the intervention displayed more benign processing and lower rates of aggressive behavior after the intervention.

Lochman and Wells (2004) developed a social–cognitive skills-training program, called the Coping Power Program, that has been found to have positive effects in reducing aggressive behavior, as rated by school teachers, that persisted into the following school year. Ross and Ross (1998) found positive effects in preventing reoffending for their program aimed at helping youths to stop and think about social problems, consider alternative strategies, and consider consequences of their actions. Finally, Kazdin (2003) developed an approach that he calls Problem-Solving Skills Training, which has produced success in reducing aggressive behavior that lasts over 12 months in both home and school settings, in at least five replicated random-assignment studies.

TOWARD A MODEL OF THE TRANSACTIONAL INFLUENCE OF RESPONSE EVALUATION AND DEVELOPMENT AND AGGRESSIVE BEHAVIOR

Recently, Fontaine et al. (2008) conducted a five-wave study of the longitudinal reciprocal influence of adolescents' aggressive RED processing and externalizing conduct problems from Grades 7 to 12. Of interest was how adolescents' RED processing of aggressive response options and their enactments of antisocial behavior inform and impact each other across adolescent development. It was hypothesized that despite substantial evidence that antisocial functioning is highly stable during adolescence (Adams, Bukowski, & Bagwell, 2005; Loeber & Hay, 1997; Olweus, 1979), youths' evaluations and decisions about acting aggressively play an important role in dynamic behavioral change across this developmental period. Specifically, three main hypotheses were tested:

1. Adolescent RED processing of aggressive behaviors mediates change in early adolescence (i.e., from Time 1 to Time 3). Adolescents' antisocial enactments are experienced in ways that inform subsequent evaluative processing of aggressive behavioral options, that, in turn guides future antisocial behavior.
2. Adolescent antisocial behavior mediates change in aggressive RED in middle adolescence (i.e., from Time 2 to Time 4). Adolescents evaluate aggressive behaviors in ways that inform their antisocial behavioral enactments, which, in turn, are experienced in ways that influence future RED processing.
3. As an extension of the first hypothesis, aggressive RED continues to mediate the development of antisocial behavior through late adolescence (i.e., from Time 3 to Time 5).

This study was conceptualized as a developmental reciprocation of individual systems (social cognition and behavior) by which one's evaluative judgments about aggressive retaliation affects his or her subsequent behavioral enactments, and in turn, one's experiences of these enactments serves to guide how behavioral options are processed and evaluated in the future. It should be noted that this examination did not test transactions between the child and the context. Although adolescents' antisocial behaviors were assessed through both self-report and parent report, environmental responsivity to adolescents' antisocial behavior was not measured. Still, this study may serve as a critical step toward understanding the role of response decision making in the transactional development of antisocial behavior in adolescence.

This investigation was conducted as part of the Child Development Project (Dodge, Bates, & Pettit, 1990), a longitudinal study of the development of conduct problem behaviors in a community sample. In 1987 and 1988, two cohorts of children from three geographic sites (Bloomington, Indiana, and Knoxville and Nashville, Tennessee) were recruited as they preregistered for public kindergarten. The first assessment included 585 primarily European American lower- to middle-class families. The participants subsequently have been followed through early adulthood.

Adolescents' externalizing problems were assessed through the Youth Self-Report and Child Behavior Checklist , administered at Times 1, 3, and 5 (Grades 7, 9/10, and 12) of the study (Achenbach, 1991a, 1991b, 1991c). In addition, RED was assessed at Times 2 and 4 (Grades 8 and 11) as part of a larger SIP assessment involving video-recorded social interactions portraying adolescents in everyday decision-making situations. Participants viewed six vignettes and were asked to imagine themselves as the respondent in each situation. Vignettes were shown in two segments. In the first segment, the respondent is ambiguously provoked by a peer or authority figure. For example, in one vignette that is set at the school track, a new student (Girl 1; respondent) approaches a peer (Girl 2; ambiguous provocateur) and asks her if she ran cross-country the previous year. Girl 2 answers, "Sure did. Are you one of the new girls?" Girl 1 replies, "Yeah. How are the workouts?" With neutral affect and intonation, Girl 2 responds, "They're tough if you're not in shape." The second segment of each vignette shows the respondent retaliating aggressively toward the ambiguous provocateur. For example, in the school track vignette, Girl 1 retaliates by snapping, "I don't think I'll have any problems, but I doubt you'll make it." Following the second segment of each vignette, participants answered evaluative questions about the aggressive retaliation that is portrayed. The questions were designed to represent qualitatively distinct domains of RED: (a) "How easy would it be for you to act like this?" (response efficacy); (b) "If you acted this way, how would you feel about yourself?" (emotional outcome expectancy); and (c) "How much would other people like you if they saw you acting like this?" (social outcome expectancy). Questions were answered according to corresponding 5-point continuous scales. These video stimuli have been used successfully in numerous other empirical investigations (e.g., Dodge et al., 2003; Fontaine et al., 2002).

Structural equation modeling was used to examine the hypotheses for various reasons. Item indicators of externalizing problems were created to construct an antisocial latent variable at Times 1, 3, and 5. With respect to RED, item indicators were based on average scores of RED domain questions across vignettes at Times 2 and 4. All hypotheses were empirically supported, and the overall fit of the model was discerned to be satisfactory. The main findings of the study were as follows:

1. Measures of antisocial behavior across adolescence (from Time 1 to Time 3 to Time 5) and aggressive RED from Time 2 to Time 4 were highly stable.

2. Antisocial behavior was predicted by prior social–cognitive processing, and aggressive RED was predicted by earlier externalizing conduct problems.
3. Aggressive RED mediated change in antisocial behavior in early adolescence from Time 1 to Time 3 and also from Time 3 to Time 5.
4. Antisocial behavior mediated aggressive RED processing from Time 2 to Time 4.

These findings supported our hypothesized model of aggressive RED and antisocial behavior, which was conceptualized as reciprocally influential from early to late adolescence. An individual systems model posits that youths' antisocial behavioral enactments influence their future processing and decision making in interpersonal situations; in turn, aggressive RED proximally affects future antisocial conduct. For example, the youth who engages in antisocial behavior toward a peer may feel more capable and satisfied with the outcome and, as a result, may evaluate aggressive response options more favorably in the future; in turn, he or she may be more likely to select and enact antisocial behaviors when responding in subsequent peer interactions.

This study is one of very few to find evidence that social cognition plays a role in actual developmental change in antisocial behavior. This is particularly noteworthy given the high stability of antisocial behavior that has been repeatedly demonstrated—a finding that was replicated in the current study. Also important to note is that not only was behavior predicted by social–cognitive processing but social cognition was also predicted by behavior. The processes by which youths' social behaviors influence subsequent social–cognitive processing are complex and, to date, not adequately investigated or understood. Findings of this study suggest that antisocial behavior may be experienced in ways that affect developmental change in online social–cognitive processing.

FUTURE DIRECTIONS FOR RESEARCH

We have argued that greater attention must be paid to the role of evaluative decision making in the development of antisocial and criminal behavior in adolescence. As compared with earlier developmental periods, adolescents have significantly greater cognitive skills with respect to considering and associating risks and possible outcomes with alternative behavioral options (Keating, 2004; Steinberg, 2005). Increasingly, these advances in executive function and behavioral decision making have been considered in light of adolescent patterns of serious antisocial behavior (e.g., Fried & Reppucci, 2001; Miller & Byrnes, 1997; Steinberg & Cauffman, 1996). Understanding developmental changes and differences in adolescent executive function has important implications for legal and policy considerations surrounding criminal culpability. In addition to basic science interests, our findings are relevant for the interdisciplinary study of psychology and law. For example, one crucial issue has to do with at what point adolescent reasoning, evaluative judgment, and decision making begin to approximate those capabilities in adults.

In this vein, it should be recognized that there remain important developmental questions about RED, and future research should examine the degree to which this level of processing contributes to antisocial development that extends beyond adolescence into adulthood. In addition to extending the developmental study of RED beyond adolescence, it is equally critical for future research to test a more comprehensive RED model (see Fontaine & Dodge, 2006). The model of RED that was assessed in the Fontaine et al. (2008) study was based on an early conceptualization of the response decision step (see Crick & Dodge, 1994). As such, it did not include multiple processes that are believed to be of potential importance to adolescents' response evaluation and decision making. Fontaine and Dodge stressed the need to consider outcome valuations in addition to, and in light of, outcome expectancies as well as response valuations that focus on the social and moral acceptability of the behavior being considered. Future studies need to incorporate these and additional aspects of RED, which there is theoretical and empirical reason to believe are potentially active during adolescents' RED processing.

Likewise, this examination further points to the importance of studying the role of youth behavior in the development of social cognition. The need for more longitudinal and experimental research that looks at this developmental sequence is substantial. Studies that focus on aspects of SIP as outcomes are important not only to a better understanding of how social–cognitive processing changes across development but also to a fuller comprehension of the dynamic nature of social cognition and behavior.

Another critical issue has to do with the appropriate unit of analysis in research on the bidirectionality of individual systems and individual–environment transactions. Future study may be conducted such that decision making and behavior are measured more frequently across shorter time intervals. The Fontaine et al. (2008) investigation was limited to examining two mediations of change in antisocial behavior by RED and one mediation of RED by antisocial behavior. An improvement would be marked by scheduling RED assessments at Grades 6, 8, 10, and 12, thus allowing for tests of two additional mediations of RED by antisocial behavior (one at early adolescence, from Grade 6 to Grade 8, and one at late adolescence, from Grade 10 to Grade 12). This design is consistent with current RED formulations that posit that evaluative decision processing and antisocial behavior continue to reciprocally influence each other across adolescent development (Fontaine & Dodge, 2006).

Also important to advancing research on the transactional development of antisocial behavior is extending our work to include repeated measurements of environmental responsivity (or social exchange, as depicted in Figure 7.1). As recognized, this component was not included in the Fontaine et al. (2008) study. As a result, the role of RED in adolescents' individual–environment transactions remains unclear.

Another important issue with respect to measurement has to do with contemporaneous versus longitudinal reciprocal influence. Whereas contemporaneous reciprocal-influence models represent relations among variables that are measured at the same time across multiple waves (e.g., Luskin, 1990), longitudinal reciprocal-influence models are estimated on the basis of variables that are repeatedly measured across unique sets of developmental time points such that no two variables are assessed simultaneously (or at the same time point; e.g., Fontaine et al., 2008; Huesmann & Guerra, 1997). It may be argued that developmental phenomena are more appropriately investigated and analyzed through longitudinal reciprocal-influence modeling because change, by its nature, necessitates the passing of time. This reasoning holds with both studies of dynamic macroprocess (the interaction of social–cognitive processing and behavior) as well as microprocess (individual online response judgments and behavioral enactments that are situation specific) exchanges. For a discussion of statistical approaches to overcoming this difficulty see Gonzalez's work (chap. 12, this volume) and Gershoff, Aber, and Clements's contribution (chap. 11, this volume).

At the theoretical level, it is essential to the study of transactional development of antisocial behavior to account for the role of SIP. Although it may be that different steps of SIP play more prominent roles in antisocial behavioral functioning at different stages of youth development (see Lansford et al., 2006), the plausibility of the mediational function of SIP remains the same. In fact, SIP may serve to mediate not only the effects of environment on how one responds behaviorally but also the process by which one's behavior affects his or her social world. If this theoretical proposition is valid—and there is an abundance of empirical research that supports such a perspective—then it must follow that neither path of a transactional reciprocation may be fully explained without understanding the mediating processing mechanisms.

REFERENCES

Achenbach, T. M. (1991a). *Integrative guide for the 1991 CBCL/4-18, YSR, and TRF profiles*. Burlington: University of Vermont Department of Psychiatry.

Achenbach, T. M. (1991b). *Manual for the Child Behavior Checklist/4-18 and 1991 profile*. Burlington: University of Vermont Department of Psychiatry.

Achenbach, T. M. (1991c). *Manual for the Youth Self-Report and 1991 profile*. Burlington: University of Vermont Department of Psychiatry.

Adams, R. E., Bukowski, W. M., & Bagwell, C. (2005). Stability of aggression during early adolescence as moderated by reciprocated friendship status and friend's aggression. *International Journal of Behavioral Development, 29*, 139–145.

Asarnow, J. R., & Callan, J. W. (1985). Boys with peer adjustment problems: Social–cognitive processes. *Journal of Consulting and Clinical Psychology, 53*, 80–87.

Bandura, A. (1977). Self-efficacy: Toward a unifying theory of behavioral change. *Psychological Review, 84*, 191–215.

Cicchetti, D., Toth, S. L., Maughan, A. (2000). An ecological–transactional model of child maltreatment. In A. Sameroff, M. Lewis, & S. Miller (Eds.), *Handbook of developmental psychopathology* (2nd ed., pp. 689–722). New York: Plenum Press.

Choudhury, S., Blakemore, S. J., & Charman, T. (2006). Social cognitive development during adolescence. *Social Cognitive and Affective Neuroscience, 1*, 165–174.

Cobb, N. J. (1992). *Adolescence: Continuity, change, and diversity*. Mountain View, CA: Mayfield Publishing.

Crick, N. R., & Dodge, K. A. (1994). A review and reformulation of social information-processing mechanisms in children's social adjustment. *Psychological Bulletin, 115*, 74–101.

Crick, N. R., & Ladd, G. (1990). Children's perceptions of the consequences of aggressive behavior: Do the ends justify being mean? *Developmental Psychology, 26*, 612–620.

Deluty, R. H. (1983). Children's evaluations of aggressive, assertive, and submissive responses. *Journal of Clinical Child Psychology, 12*, 124–129.

Dodge, K. A., Bates, J. E., & Pettit, G. S. (1990). Mechanisms in the cycle of violence. *Science, 250*, 1678–1683.

Dodge, K. A., Lansford, J. E., Burks, V. S., Bates, J. E., Pettit, G. S., Fontaine, R., et al. (2003). Peer rejection and social information-processing factors in the development of aggressive behavior problems in children. *Child Development, 74*, 374–393.

Dodge, K. A., Lochman, J. E., Harnish, J. D., Bates, J. E., & Pettit, G. S. (1997). Reactive and proactive aggression in school children and psychiatrically impaired chronically assaultive youth. *Journal of Abnormal Psychology, 106*, 37–51.

Dodge, K. A., & Newman, J. P. (1981). Biased decision-making processes in aggressive boys. *Journal of Abnormal Psychology, 90*, 375–379.

Dodge, K. A., & Pettit, G. S. (2003). A biopsychosocial model of the development of chronic conduct problems in adolescence. *Developmental Psychology, 39*, 349–371.

Dodge, K. A., & Price, J. M. (1994). On the relation between social information processing and socially competent behavior in early school-aged children. *Child Development, 65*, 1385–1398.

Erdley, C. A., & Asher, S. R. (1996). Children's social goals and self-efficacy perceptions as influences on their responses to ambiguous provocation. *Child Development, 67*, 1329–1344.

Feather, N. T. (1982). Expectancy-value approaches: Present status and future directions. In N. T. Feather (Ed.), *Expectations and actions: Expectancy-value models in psychology* (pp. 395–420). Hillsdale, NJ: Erlbaum.

Fontaine, R. G. (2006a). Applying systems principles to models of social information processing and aggressive behavior in youth. *Aggression and Violent Behavior, 11*, 64–76.

Fontaine, R. G. (2006b). Evaluative behavioral judgments and instrumental antisocial behaviors in children and adolescents. *Clinical Psychology Review, 26*, 956–967.

Fontaine, R. G. (2007a, March). *Social information processing, response evaluation and decision, and antisocial behavior: Developmental science and implications for criminal law*. Paper presented to the Department of Psychology, Sapienza University of Rome, Italy.

Fontaine, R. G. (2007b). Toward a conceptual framework of instrumental antisocial decision-making and behavior in youth. *Clinical Psychology Review, 27*, 655–675.

Fontaine, R. G., & Burks, V. S. (1999, April). *Dimensions of response-decision style and externalizing behavior in adolescence*. Poster session presented at the meeting of the Society for Research in Child Development, Albuquerque, NM.

Fontaine, R. G., Burks, V. S., & Dodge, K. A. (1998, March). *The mediating effect of sociomoral judgments about aggression on the relation between hostile attributional style and antisocial conduct*. Paper presented at the meeting of the Conference on Human Development, Mobile, AL.

Fontaine, R. G., Burks, V. S., & Dodge, K. A. (2002). Response decision processes and externalizing behavior problems in adolescents. *Development and Psychopathology, 14*, 107–122.

Fontaine, R. G., & Dodge, K. A. (2006). Real-time decision making and aggressive behavior in youth: A heuristic model of response evaluation and decision (RED). *Aggressive Behavior, 32*, 604–624.

Fontaine, R. G., Yang, C., Dodge, K. A., Bates, J. E., & Pettit, G. S. (2008). Testing an individual systems model of response evaluation and decision (RED) and antisocial behavior across adolescence. *Child Development, 79*, 462–475.

Fried, C. S., & Reppucci, N. D. (2001). Criminal decision making: The development of adolescent judgment, criminal responsibility, and culpability. *Law and Human Behavior, 25*, 45–61.

Granic, I., & Patterson, G. R. (2006). Toward a comprehensive model of antisocial development: A dynamic systems approach. *Psychological Review, 113*, 101–131.

Guerra, N. G., Huesmann, L. R., & Hanish, L. (1995). The role of normative beliefs in children's social behavior. In N. Eisenberg (Ed.), *Social development* (pp. 140–158). Thousand Oaks, CA: Sage.

Guerra, N. G., & Slaby, R. G. (1989). Evaluative factors in social problem solving by aggressive boys. *Journal of Abnormal Child Psychology, 17*, 277–289.

Guerra, N. G., & Slaby, R. G. (1990). Cognitive mediators of aggression in adolescent offenders: II. Intervention. *Developmental Psychology, 26*, 269–277.

Hart, D., & Yates, M. (1997). The interrelation of self and identity in adolescence: A developmental account. In R. Vasta (Ed.), *Annals of child development: A research annual* (Vol. 12, pp. 207–243). Philadelphia: Kingsley Publishers.

Harter, S., Bresnick, S., Bouchey, H. A., & Whitesell, N. R. (1997). The development of multiple role-related selves during adolescence. *Development and Psychopathology*, 9, 835–853.

Hollingshead, A. A. (1979). *Four-factor index of social status*. Unpublished manuscript, Yale University, New Haven, CT.

Hudley, C. A., & Graham, S. (1993). An attributional intervention to reduce peer-directed aggression among African-American boys. *Child Development, 64*, 124–138.

Huesmann, L. R. (1998). The role of social information processing and cognitive schema in the acquisition and maintenance of habitual aggressive behavior. In R. G. Geen & E. Donnerstein (Eds.), *Human aggression: Theories, research, and implications for social policy* (pp. 73–109). San Diego, CA: Academic Press.

Huesmann, L. R., & Guerra, N. G. (1997). Children's normative beliefs about aggression and aggressive behavior. *Journal of Personality and Social Psychology, 72*, 408–419.

Huesmann, L. R., & Moise, J. F. (1998). The stability and continuity of aggression from early childhood to young adulthood. In D. J. Flannery & C. R. Huff (Eds.), *Youth violence: Prevention, intervention, and social policy* (pp. 73–95). Washington, DC: American Psychiatric Press.

Kazdin, A. E. (2003). Problem-solving skills training and parent management training for conduct disorder. In A. E. Kazdin & J. R. Weisz (Eds.), *Evidence-based psychotherapies for children and adolescents* (pp. 241–262). New York: Guilford Press.

Kalogerakis, M. G. (2003). Adolescent violence in America: A historical perspective. *Adolescent Psychiatry, 27*, 3–28.

Keating, D. P. (2004). Cognitive and brain development. In R. J. Lerner & L. D. Steinberg (Eds.), *Handbook of adolescent psychology* (2nd ed., pp. 45–84). New York: Wiley.

Killeen, M. R., & Forehand, R. (1998). A transactional model of adolescent self-esteem. *Journal of Family Psychology, 12*, 132–148.

Lansford, J. E., Malone, P. S., Dodge, K. A., Crozier, J. C., Pettit, G. S., & Bates, J. E. (2006). A 12-year prospective study of patterns of social information processing problems and externalizing behaviors. *Journal of Abnormal Child Psychology, 34*, 709–718.

Lochman, J. E., & Wells, K. C. (2004). The coping power program for preadolescent aggressive boys and their parents: Outcome effects at the 1-year follow-up. *Journal of Consulting and Clinical Psychology, 72*, 571–578.

Loeber, R., & Hay, D. (1997). Key issues in the development of aggression and violence from childhood to early adulthood. *Annual Review of Psychology, 48*, 371–410.

Luskin, R. C. (1990). Explaining political sophistication. *Political Behavior, 12*, 331–361.

Miller, D., & Byrnes, J. (1997). The role of contextual and personal factors in children's risk taking. *Developmental Psychology, 33*, 814–823.

Olweus, D. (1979). Stability of aggressive patterns in males: A review. *Psychological Bulletin, 86*, 852–875.

Perry, D. G., Perry, L. C., & Boldizar, J. P. (1990). Learning of aggression. In M. Lewis & S. Miller (Eds.), *Handbook of developmental psychopathology* (pp. 135–145). New York: Plenum Press.

Rabiner, D. L., & Coie, J. D. (1989). The effect of expectancy inductions on rejected children's acceptance by unfamiliar peers. *Developmental Psychology, 25*, 450–457.

Roesch-Ely, D., Spitzer, M., & Weisbrod, M. (2003). Cognitive inhibition and thought disorder in schizophrenia. *Psychopathology, 36*, 23–32.

Ross, R. R., & Ross, B. D. (1998). Delinquency prevention through cognitive training. *New Education, 10*, 70–75.

Sameroff, A. J. (2000). Developmental systems and psychopathology. *Development and Psychopathology*, 12, 297–312.

Sameroff, A. J., & Chandler, M. J. (1975). Reproductive risk and the continuum of caretaking casualty. In F. D. Horowitz (Ed.), *Review of child development research* (Vol. 4, pp. 187–244). Chicago: The University of Chicago Press.

Sameroff, A. J., & MacKenzie, M. J. (2003). Research strategies for capturing transactional models of development: The limits of the possible. *Development and Psychopathology, 15*, 613–640.

Shaw, D. S., & Winslow, E. B. (1997). Precursors and correlates of antisocial behavior from infancy to preschool. In D. Stoff, J. Breiling, & J. D. Maser (Eds.), *Handbook of antisocial behavior* (pp. 148–158). New York: Wiley.

Slaby, R. G., & Guerra, N. G. (1988). Cognitive mediators of aggression in adolescent offenders: 1. Assessment. *Developmental Psychology, 24*, 580–588.

Smithmyer, C. M., Hubbard, J. A., & Simons, R. F. (2000). Proactive and reactive aggression in delinquent adolescents: Relations to aggression outcomes expectancies. *Journal of Clinical Child Psychology, 29*, 86–93.

Steinberg, L. (2005). Cognitive and affective development in adolescence. *Trends in Cognitive Sciences, 9*, 69–74.

Steinberg, L., & Cauffman, E. (1996). Maturity of judgment in adolescence: Psychosocial factors in adolescent decision making. *Law and Human Behavior, 20*, 249–272.

Steinberg, L., & Scott, E. S. (2003). Less guilty by reason of adolescence: Developmental immaturity, diminished responsibility, and the juvenile death penalty. *American Psychologist, 58*, 1009–1018

III

SOCIALIZATION AND EDUCATION

8

TOWARD A MODEL OF CULTURE↔PARENT↔CHILD TRANSACTIONS

MARC H. BORNSTEIN

Transactional processes in parent–child dyads are an everyday affair, but their form and function vary dramatically with culture. After laying some groundwork about the basic structure of transactions and invoking some common examples from U.S. American culture, in this chapter I consider illustrations of culture-specific and culture-universal parent–child transactions in development. A theoretical discussion of form and function with regard to transaction in cultural contexts follows, and the chapter concludes with a discussion of the unique contributions a cultural perspective offers to the theoretical notion of developmental transactions.

This chapter was authored by an employee of the U.S. government as part of official duty and is considered to be in the public domain. Any views expressed herein do not necessarily represent the views of the United States government, and the author's participation in the work is not meant to serve as an official endorsement. Preparation was supported by the Intramural Research Program of the National Institutes of Health and the Eunice Kennedy Shriver National Institute of Child Health and Development. I refer to different cultural studies by way of illustration. Some come from others' research programs, some from my own.

CONSISTENCY IN CHILD, PARENT, AND CULTURE

Human beings vary dramatically on every psychological construct, structure, function, or process (in what follows, I use P to stand for any such entity). Variation in Ps appears in normal distributions in the population, and individual differences in some Ps are fleeting, whereas others are relatively stable through time. Psychological science is centrally concerned with assessment of consistency in P. Furthermore, it is desirable to distinguish between consistency among individuals—individual order consistency (stability or instability) through time—and consistency in a group—group mean level consistency (continuity or discontinuity) through time. Order consistency and mean consistency are independent constructs. Conceptually and empirically, the two can exist simultaneously, and both have been reported in longitudinal studies of different Ps. An individual with a stable P will display a relatively high level of P at one point in time vis-à-vis peers and continue to display a high level of P at a later point in time, whereas other individuals display lower levels of P at both times. Individuals show instability in a given P if they do not maintain their relative standing or order in the group through time. A group with a continuous P will display a mean level of P at one point in time and continue to display that level of P at a later point in time. A group will show discontinuity in a given P if it does not maintain its mean level through time.

Consistency in development depends on the P (some Ps may be more consistent than others); the age of the individual when P is assessed (a P may not be consistent at one age or stage but may stabilize at another); the frequency or duration of the P assessment (more data points normally yield more stable estimates); the temporal interval between assessments of P (the shorter the interassessment interval, normally the greater the consistency of P); and whether assessments of P are made across the same or different contexts (normally the former improves consistency and the latter attenuates consistency).

Psychological consistency is important for several reasons. First, the physiological systems of humans and other living organisms require certain consistent conditions—physical, chemical, psychological, and environmental—to survive. Second, consistency provides basic information about the overall developmental course of a given P. Third, it is generally assumed that to be meaningful, a P should show substantial consistency across time: A major predictor of P at a given age is P at an earlier age. Fourth, psychological consistency affects the environment: Interactants often adjust to a consistent P in an individual (which may contribute to the consistency).

Developmental science also emphasizes change through time. Change arises because it is programmed into the genome, because P is plastic to psychosocial experience, and because people engage in normative life tasks and roles. Although genetic programming is usually considered the source of continuity, its expression constantly changes as a function of external as well as internal experiences. The life-span perspective in psychology asserts that human beings are open systems, and the plastic nature of psychological functioning (ψ) ensures that people exhibit both consistency and inconsistency in many Ps throughout the life course.

Development is governed by genetic and biological factors in combination with environmental influences and psychosocial experiences. Thus, stability and change in any P are attributable to genetic or biological factors in the individual and emerge through the individual's transactions with the environment.

Consistency in Children

Children change, yet remain the same through time. Children carry with them in their development their cognitive styles, personality, temperaments, and other aspects of their uniqueness. This is so-called intraindividual consistency. The developmental literature supports both consistency and inconsistency in temperament as well as cognitive functioning in childhood (Bornstein et al., 2006).

Consistency in Parenting

Parents too are consistent and inconsistent through time in some nontrivial degree. Consistency in parenting is of two kinds: (a) intraindividual consistency, that is, stability of standing in the group or group mean level continuity across time, and (b) transgenerational consistency, that is, consistency from parent to child (as eventual parent) through time. There is ample evidence of both in the literature.

Intraindividual Consistency

Holden and Miller (1999) meta-analyzed the existing literature and found that parental cognitions and practices are relatively consistent across time, children, and situations, but parenting is also characterized by change, which the authors attributed to the adaptability of parents to varying demands of different contexts. Dallaire and Weinraub (2005) reported on a diverse sample of 1,364 families enrolled in the Study of Early Child Care at 10 sites across the United States. Parenting behaviors were rated annually over a 6-year period using laboratory observations and home visits. In the study, 893 participants had complete data from each time point. Sensitive and stimulating parenting displayed considerable relative (stability) and mean level (continuity) consistency over the first 6 years of life, whereas negative aspects of parenting, such as detachment and negative regard, did not.

Through intergenerational transmission, by interlocked genetic and experiential pathways, purposefully or unintentionally, one generation also influences the parenting beliefs and behaviors of the next (van IJzendoorn, 1992). Fraiberg and her colleagues (Fraiberg, Adelson, & Shapiro, 2003) famously referred to these influences as "ghosts in the nursery." A parent's experiences with his or her own parents have continuing effects on his or her own parenting (Smith & Drew, 2002). Examples abound in the literature. Ruoppila (1991) found significant correlations between grandparental and parental child rearing in a Finnish sample. Vermulst, de Brock, and van Zutphen (1991) examined parental functioning across generations in a Dutch sample of grandmother–mother dyads: Approximately one third of the variation in mothers' parental functioning could be explained in terms of parental functioning of the grandmother. The use of physically aggressive and punitive techniques in the grandparent–parent generation predicts similar behavior in the parent–grandchild generation and antisocial behavior in grandchildren (Farrington, 1994). Marital violence in the family of origin tends to repeat in the successive generations (Stith, Rosen, & Middleton, 2000). When parents abuse their children, children are at risk of repeating abusive patterns as adults with their own children (Newcomb & Locke, 2001). Meta-analyses find that adult children whose parents divorce are more likely to end their own marriages in divorce (Chase-Lansdale, Cherlin, & Kiernan, 1995). Maritally dissatisfied couples are more likely to have had unhappily married parents (Amato & Booth, 2001). All that said, parenting undergoes frequent adjustments and is itself a product of multiple transactional processes.

Consistency in Environment

The environment of human development has both physical features, for example, climate and designed artifacts, and social features, for example, culture, that also have a dynamic aspect over time. This chapter is concerned principally with cultural aspects of the environment. Cultures consist of distinctive patterns of affiliation as well as norms, ideas, values, and assumptions about life that are shared by a group of people and that guide and regulate specific behaviors and inculcate valued and adaptive competencies. Culture constitutes an abstract entity of learned meanings and shared information transmitted from one generation to the next through social interaction. The concept of culture is therefore frequently used as a means of understanding relations between environments on the one hand and individual psychological ψs on the other. Every psychological construct, structure, function, and process has cultural continuo and overtones. Physical and social environments are hardly static entities; rather, they are dynamic systems that are constantly in the process of reconstruction and renegotiation in the context of individual lives. Cultures change but are also consistent through time.

TRANSACTION IN THEORY AND TRANSACTIONS IN REALITY

Figure 8.1 shows these consistencies and changes through time for child, parent, and environment. The figure also shows concurrent interactions and temporal transactions between levels.

Parent–Child and Child–Parent Transactions

Child and parent bring distinctive characteristics to, but each also changes as a result of, interactions with one another; both parent and child then enter the next interaction as changed individuals (see Figure 8.1). Of course, change can be expected to attenuate consistency. Transaction asserts that the characteristics of an individual shape his or her experiences and, reciprocally, experiences shape the characteristics of the individual through time. Development is a function of the individual and the individual's environment and not of either alone.

For a long time, parenting was identified with unidirectional socialization. Children's actions, and more subtly their perceptions or interpretations of parents, are equally meaningful to parents' actual cognitions and practices

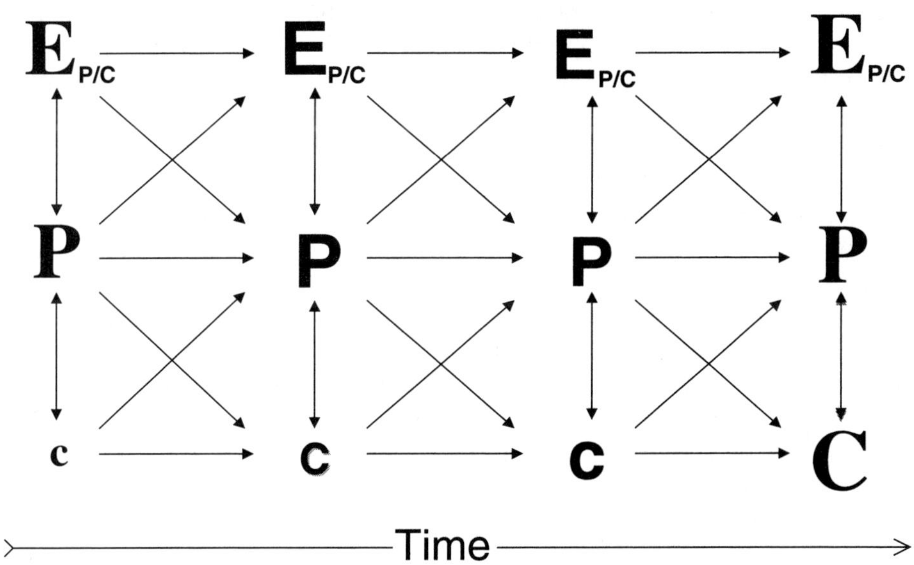

Figure 8.1. Reciprocal relations between environment (E) and parent (P) and between parent (P) and child (C). Different fonts for E, P, and C indicate inconsistencies and variability across time.

in influencing development (Noller, Seth-Smith, Bouma, & Schweitzer, 1992). Children actively select, modify, and create their own environments, including their parenting (Bell, 1968; Scarr & Kidd, 1983). Children influence some experiences they will be exposed to, and they interpret and appraise those experiences and so, in some degree, determine how those experiences will affect them (Scarr & McCartney, 1983). What children imitate in others, what they remember, and how they process what is remembered all depend on their cognition, personality, and level of development. Piaget's interactionist theorizing, for example, strongly suggested that children use parental input rather than parental input overdetermining what children learn.

Two-Term Transactions

When child affects parent (*child effects*) or parent affects child (*parent effects*), two-term transactions are in evidence. A reciprocal effects or transactional model asserts that parents not only affect their child but the child also affects parents. Children's achieving the ability to stand upright and walk alters the nature and quality of the parenting they receive (Biringen, Emde, Campos, & Appelbaum, 1995). With each child advance, parenting changes in some corresponding ways.

Parent Effects

A central assumption of all socialization theories is that even though socialization and resocialization can occur at any point in the life cycle and that parent–child interactions robustly influence children, childhood is a particularly plastic period when enduring social skills, personality attributes, and cultural values are inculcated. Parents' cognitions and practices influence child development through different paths, even if so far developmental science has advanced only a handful of processes by which parenting influences child development: psychodynamic, learning, and cognitive. For example, Vygotsky (1978) contended that as a central feature of the transactional perspective (see chap. 9, this volume), the more advanced partner (the parent) raises the level of competence or performance in the less advanced partner (the child). A common assumption in parenting is that the overall level of parental involvement or stimulation affects the child's overall level of development (Maccoby & Martin, 1983). Increasing evidence suggests, however, that specific and more differentiated processes explain parenting effects. For example, the effectiveness of a given experience may depend on its process pathway (Bornstein, 2006).

Child Effects

Transactional theoretical formulations acknowledge the role of the child in interaction with the parent. Bell (1968; Bell & Harper, 1977) emphasized the key role that child effects play in socialization. An experimental study by Anderson, Lytton, and Romney (1986) paired conduct-disordered boys with mothers of conduct-disordered boys and with mothers of normal boys (see chap. 6, this volume). Conduct-disordered boys elicited negative parenting practices from both sets of mothers. Clearly, in this experimental setting (which controls for shared genetic effects), behavioral characteristics of the child contribute to the type of parenting displayed. Likewise, stage of child development exerts pervasive control over parental behavior. Cross-cultural study shows that mothers of younger children use more affect-laden speech, but that as children achieve more sophisticated levels of motor exploration and cognitive comprehension, mothers increasingly orient, comment, and prepare children for the world outside the dyad by increasing amounts of information in their speech (Bornstein et al., 1992). Over the second half of the first year of life, as infants achieve more sophisticated levels of motor exploration and cognitive comprehension, their mothers—from the United States, Argentina, and Japan alike—used language increasingly to reference the environment to orient, comment, and prepare infants for the world outside the dyad. That is, mothers expect that their infants need to be directed more, expect that their infants know more or better comprehend their questions, and expect that they can and should give their infants more information about the infants themselves, their mothers, and the environment.

Children (even in the same family) engender different interactions not only because they differ in age, cognitive level, personality characteristics, or gender but also because different children interpret interactions with their parents differently.

Three-Term Transactions

Three-term transactions define dynamics in which the child changes the parent and is in turn changed by the changed parent, or the parent changes the child who in turn changes the parent. Demonstrators of three-term transactions are less frequent in empirical literature, although they must be common in everyday life. Thomas, Chess, and Birch (1968) elucidated such a transactional developmental path for children with difficult temperaments. These children stimulated maladaptive parenting that led to their own later behavioral disturbance. For children who did not have this transaction occur (whose parents were not negatively reactive to the temperaments of their children), no such pathway to behavioral deviance was found. When parents respond to an irritable, difficult-to-soothe infant with hostility and anger, the infant is placed at even greater risk of later maladjustment than by difficult temperament alone. Bates, Pettit, and Dodge (1995), found that harsh parental reactions to 4-year-olds identified in infancy as temperamentally difficult enhanced the effect of temperamental difficulty on externalizing behavior in adolescence. Similarly, Belsky, Hsieh, and Crnic (1998) observed that negative and intrusive mothering predicted externalizing behavior only in 3-year-olds judged at 1 year of age to be temperamentally difficult. Clearly, favorable child outcomes are not fostered by harsh parental reactions to temperamentally challenging infants. Gandour (1989) observed 15-month-old toddlers' exploratory activity at home as well as their mothers' efforts to guide and stimulate them. Toddlers with low activity levels whose mothers provided more stimulation explored more than equally low-activity toddlers with understimulating mothers.

By virtue of their unique characteristics and propensities—state of arousal, perceptual awareness, cognitive status, emotional expressiveness, and individuality of temperament and personality—children and parents actively contribute, through their interactions, to produce one another's development.

Culture–Parent and Parent–Culture Transactions

Just as there are reciprocal relations between child and parent as indicated in the bottom two rows of Figure 8.1, there are reciprocal relations between environment and parent, as indicated in the top two rows of Figure 8.1. Parenting is culturally organized. Parents structure interactions with their children in everyday situations using strategies geared to actualize socialization goals that are emphasized by their cultural value system. In this sense, parents are culture bearers. Not yet well worked out are functional and theoretical connections between culture and parenting. How do components of culture relate to parenting attitudes and actions? Much more work needs to be accomplished linking parenting to culture, just as more work needs to link culture to parenting. Nonetheless, *thematicity* (the repetition of the same cultural idea across mechanisms and in a variety of contexts) has special importance in culture as an organizer of behavior (Schank & Abelson, 1977). Every culture is characterized and distinguished from others by thoroughgoing, deep-seated, and consistent themes that inculcate what one needs to know to think and behave as a functioning member of the culture (Quinn & Holland, 1987). So, for example, in the United States personal choice is closely bound up with how individuals think of themselves and make sense of their lives. Personal choice is firmly rooted in principles of liberty and freedom and is an outstanding psychological construct in the literature on U.S.-born children (Deci, Ryan, & Koestner, 1999). Choice affects aspects of life at home, school, and in neighborhoods. Indeed, choice is an unspoken value that underlies children's everyday interactions with others from their first days of life. From infancy, U.S. parents view children as intentional agents and respond to infants' initiatives as though their babies are already making choices about what to say and what they want. The U.S. view of infants as intentional agents is also reflected in the cultural value that is placed on parents' responsiveness. As infants grow into toddlers, the term *terrible twos* refers to the stubborn desire children have to "do their own thing" rather than accommodate to the will of others. Parents perceive toddlers' expressions of negativism as linked to the development of autonomy and choice, both of which are strongly encouraged. Consequently, when a parent bemoans a toddler's outbursts, it is often with implicit pride, pointing out that the child wants to "do everything himself" or that "she won't listen to anything I say!" (Tamis-LeMonda & McFadden, in press). Indeed, researchers have proposed that (a limited amount of) child noncompliance is healthy and positive in that it serves an important function by supporting children's development of autonomy (Kuczynski, Kochanska, Radke-Yarrow, & Girnius-Brown, 1987). Similarly, when disciplining their older children, parents frequently use explanatory requests accompanied by reasoning (e.g., "I'd like you to do this because . . .") rather than decontextualized commands or assertions of power (e.g., "Put it away!"), perceiving the need to appeal to children's cognitive understanding of the logic behind parental management strategies. The value of choice is implicitly reflected in typical indirect parental management strategies: Appropriate behavior in children is insufficient on its own; rather, children must themselves want to engage in the appropriate behavior. Parents continue to affirm children's rights to make choices as a pathway to children's growing autonomy. Children in the United States are born into a world of choice. For the most part, children decide what to wear and eat and the toys with which to play. Children select the sports, hobbies, classes, and majors to pursue and are later expected to follow career paths that reflect their individual talents and passions. When children in the United States enter school, their teachers have received training that emphasizes the importance of autonomy and choice in their young students. Teachers voice the opinion that it is "very important to provide children with choice" and in turn frequently present children with choices such as "Do you want to play with blocks or draw pictures?" For their part, children respond to classroom opportunities by expressing their wishes, and parents celebrate their children's individuality by permitting them to choose their friends, clothes, hobbies, careers, and partners. Children in the United States are encouraged to develop into autonomous individuals who strive to attain their full potential, make their own choices, and are equal in their rights to pursue their goals (Tamis-LeMonda & McFadden, in press). In brief, the parent–child dyad is embedded in a nexus of multiple environmental contexts, and each contributes to the critical ways parents support developing characteristics in children. The result is a greater understanding of the processes and contents of parents' particular competencies and their roles as members of a culture.

What are common and characteristic cultural ideas, customs, and the like? Some surround gender. Mondschein, Adolph, and Tamis-LeMonda (2000) found that U.S. mothers of 11-month-old boys overestimated how well their boys would crawl down a sloped pathway, whereas mothers of 11-month-old girls underestimated how well their girls would do. Subsequent tests of actual crawling ability on a sloped path revealed no systematic gender differences in infant crawling. Clearly, mothers' culturally conditioned cognitions about girls and boys color their cognitions and practices. Cultural comparisons show that virtually all aspects of parenting—cognitions as well as practices—are informed by their cultural framework. Some knowledge of child development, for example, may be universal, but some is clearly culturally constructed, and parents' knowledge varies among groups with diverse cultural values and customs. Pachter and Dworkin (1997) asked mothers from minority (Puerto Rican, African American, West Indian and Caribbean) and majority cultural groups about normal ages of attainment of typical developmental milestones during the first 3 years of life: Although all responses fit within a normative developmental range, significant differences emerged across ethnic groups for more than one third of the developmental milestones tested. Mothers in rural Thailand do not believe their newborns can see (only 1.7% believe their babies can see at 1 week, 14.7% at 1 month), and often during the day they swaddle infants on their backs or in a fabric hammock so the baby can only see a narrow slit view of the ceiling or sky (Kotchabhakdi, Winichagoon, Smitasiri, Dhanamitta, & Valyasevi, 1987). Gusii and Samoan caregivers believe that it is nonsensical to talk to infants before infants themselves are capable of speech (Ochs, 1988).

More generally, socialization involves not just inculcation and the acceptance of values, standards, and customs of society but also the ability to function in an adaptive way in the larger social context. Values, standards, and customs are not simply transmitted from one generation to the next, but to some extent at least, they are reconstructed by each generation. Thus, parents are not so much purveyors of information as helpers in setting the stage for their children to become well-functioning members of their cultural group.

To show how culture moderates transactions in child effects and parenting effects, one must examine many cultures to search for different processes. To that end, one of my programs of research has both cross-cultural and intracultural components, in the sense that I collected data from (at least) two groups in many cultures. The cross-cultural aspect allows multicultural comparisons; at the same time, the intracultural aspect allows meaningful within-culture comparisons. To effect this research, I founded an international network of parenting and child development, and 15 countries participate. The locales represented are more alike than not in terms of modernity, urbanity, economics, politics, living standards, even ecology and climate. Therefore, it has been possible to recruit some roughly equivalent sociodemographic samples. Families are also typically nuclear in organization; mothers are normally the primary caregiver in the family setting; and parents share many of the same larger goals for their children, notably physical health and well-being, social adjustment and happiness, and educational achievement and economic security. Of course, substantial differences also exist, within and between groups in terms of history, beliefs, language, and values.

Cultural variation characterizes parents' attributions about their own parenting. For example, mothers from Argentina, Belgium, Italy, Israel, Japan, and the United States were asked whether being able to comfort their child successfully when the child cries was due to their parenting ability (e.g., "I am good at this"), effort (e.g., "I have tried hard"), mood (e.g., "I am in a good mood"), task difficulty (e.g., "This is easy to do"), or a child characteristic (e.g., "My child makes this easy to do"). Japanese mothers were less likely than mothers from all other nations to attribute success to their own ability and more likely to indicate that when they were successful, it was because of the child's behavior (Bornstein et al., 1998). In certain circumstances, parents tend to believe their children are behaving intentionally in one or another way when the child's behavior may in fact be developmentally typical. Thus, parents' attributions about their children help to shape parents' caregiving practices, which in turn affect their children's lives. Numerous studies have documented that Asian parents, as compared with European American parents, attribute the effort the child directs toward schoolwork as the most important contributor to the child's academic success.

CULTURAL REPRESENTATIONS: IMPLICIT MEANINGS AND OBVIOUS UNDERSTANDINGS IN CULTURAL ENVIRONMENT, PARENTING, AND CHILDHOOD

Actively or passively, to a greater or lesser degree, intentionally or unwittingly, parents pass their culture on to their offspring. Parenting is a principal reason why individuals in different cultures are who they are and often differ so from one another. What are the chief processes of enculturation? Because culture is organized information, parenting consists of mechanisms for transmitting that information and childhood for processing that information. From a dynamic transactional perspective, both parent and child select, edit, and refashion cultural information. So, minimally, enculturation involves bidirectional processes in which adult and child play transactive roles shaped by culture.

The role of culture in parent–child transactions adds a layer of complexity to the traditional transactional model. Different activities can, of course, have different meanings in different cultures. But a given activity can have the same meaning in different cultures. It is also possible for similar activities to have different meanings in different cultures or different activities to have the same meaning in different cultures. Culture is the prime context for determining associations between activity and meaning, form and function (Bornstein, 1995). Development in culture is the prime situation for examining how the meanings of activity are afforded, shaped, and acquired. Figure 8.1 shows surface relations among environment (physical, cultural), parenting, and child development. To these one must add direct child–environment

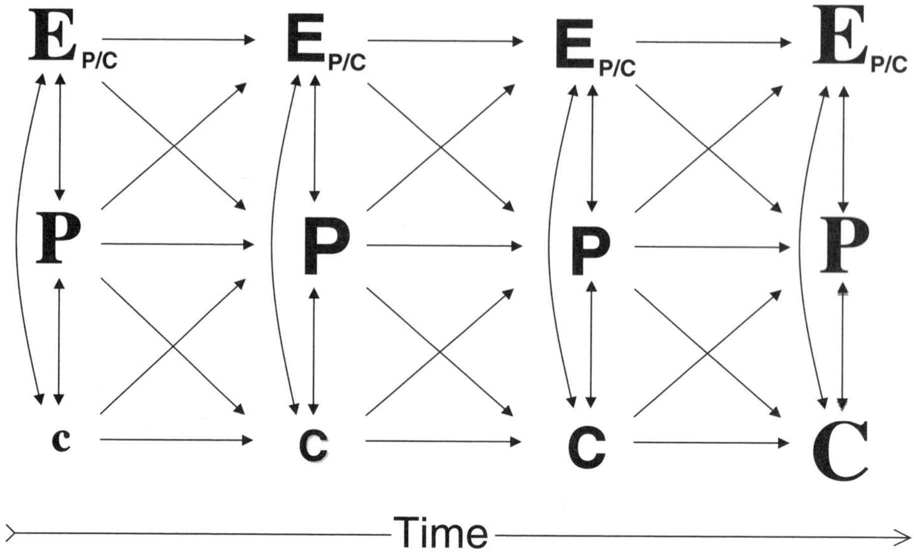

Figure 8.2. Reciprocal relations among environment (E), parent (P), and child (C). Different fonts for E, P, and C indicate inconsistencies and variability across time.

relations as depicted in Figure 8.2. However, the manifest symbols for child, parent, and environment must be interpreted as being shadowed by cultural meanings behind each of these surface behaviors as depicted in Figure 8.3.

Different Forms–Different Functions

Normally, different behaviors signal different meanings. Such a construction exemplifies cultural specificity. U.S. American parents typically wish to promote autonomy in their children, and they organize parent–child interactions to foster physical and verbal independence in children, perhaps as a reflection of or headed toward individualism and assertiveness generally associated with American culture. American mothers encourage their infants' attention to properties, objects, and events in the environment and stress functional exploratory play with their toddlers (Bornstein et al., 1992). By contrast, Japanese mothers tend to see their children as an extension of themselves and work with their children to consolidate and strengthen a mutual dependence between mother and child (*amae*), perhaps as a reflection of or headed toward collectivity and an emphasis on interpersonal sensitivity, thought to be critical in Japan. Japanese mothers encourage their infants' participation in social interactions and stress symbolic representational play with their toddlers (Bornstein et al., 1992). These interactions reflect different form–different function relations: Different kinds of interactions predict to and are shaped by individualist versus collectivist cultural tendencies.

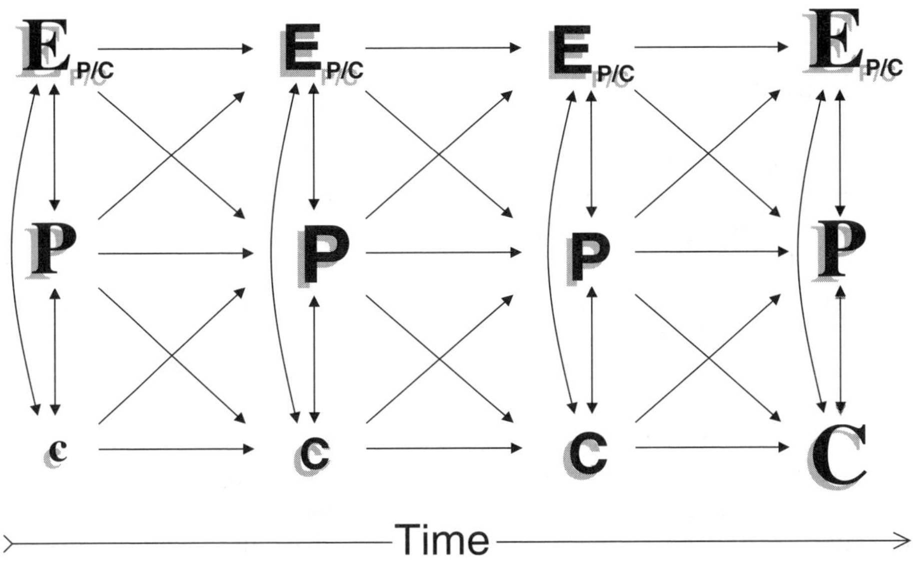

Figure 8.3. Reciprocal relations among environment (E), parent (P), and child (C). Different fonts for E, P, and C indicate inconsistencies and variability across time. Shadowing indicates cultural meanings behind surface behaviors.

Same Forms–Same Functions

It is also the case that a single form can have the same meaning in different cultures. Same form–same function relations are often thought of as cultural universals, perhaps reflective of species general tendencies. Maternal responsiveness is an example. Since Bowlby and Ainsworth, developmentalists have tended to regard prompt, contingent, and appropriate maternal responsiveness as a generally good thing. Globally, responsiveness acknowledges the integrity of the child and is thought to promote secure attachment, effectance, and a sense of self. One could expect, therefore, that some types of maternal responsiveness would be similar across cultures, perhaps all to the same effect. In fact, mothers in France, Japan, and the United States all respond to their infants' vocalizing distress predominantly (as would be expected) by nurturing (Bornstein et al., 1992). Identifying biopsychological programs that speak to species-general trends in evolutionary adaptation in human beings is far from a small matter.

Same Forms–Different Functions

Sometimes, however, the same form has different functions or meanings, depending on culture. In human mother–infant dyads, mutual eye gaze is a setting occasion for interpersonal communication and social interaction. But in other primates, prolonged gaze signals threat. When chimpanzee infants gaze at their mothers' faces, chimpanzee mothers shift their own gaze away whenever mutual eye contact is achieved. Eye gaze is learned and regulated by the social environment in chimpanzees in the very first month of life: Newborn chimps reared by human adults will engage in extended eye-to-eye contact. This dynamic is not exclusive to infrahuman cultures; culture exerts parallel effects in human behavior. In many cultures, older children are expected to look at adults directly when addressing them or when answering a question. Among some Native American cultures, however, staring directly at one's elders is a sign of disrespect; eye contact (*the evil eye*) culturally signals aggressiveness. The consequences of this stylistic difference can be telling for Native American schoolchildren who do not meet the expectations of European American schoolteachers or law enforcement officers, for example, but rather appear disinterested or (worse) disrespectful.

The attribution of *difficult temperament* in modern individualistic societies often implies a problem in the child. A difficult temperament is associated with psychiatric risk. But, infant characteristics that some adults find difficult may not pose significant challenges for others. In this sense, *difficultness* resides, not (only) in the child but in the caregiving context. Mothers are more prone to rate their firstborns *difficult*, which may derive from the fact that firstborn babies actually are more difficult or, alternatively, from the fact that first-time mothers are more ill at ease with their infants and so tend to perceive them as more difficult. Whether temperament is difficult thus depends on its fit with the culture. In an extreme example, during conditions of drought and famine in Africa, characteristics of infant difficultness were associated with survival, probably because infants with those very characteristics demanded and received more attention and nourishment. African infants who would be viewed by most Americans as difficult had better survival rates (deVries, 1984). Rudy and Grusec (2006) demonstrated that authoritarian parenting is associated with lack of warmth, feelings of inefficacy, negative cognitions about the child, and anger in Anglo Canadian parents, but not in Middle Eastern Canadian parents. Moreover, they found that children's self-esteem was more strongly predicted by these negative emotions and cognitions than it was by parental control. It is the association with negative cognition and affect that accounts for the deleterious impact of authoritarian parenting in Western cultural contexts. In non-Western cultural contexts, the same parenting style has fewer harmful outcomes (e.g., Lindahl & Malik, 1999) or is even positively associated with warmth and acceptance (Rohner & Pettengill, 1985). In these same cultures, however, when authoritarian parenting is operationalized as domineering or overprotective, it has negative effects (Herz & Gullone, 1999), just as it does in a Western cultural context.

Work in China and Canada with respect to children's behavioral inhibition, parental responses, and children's development is also illustrative. Some children are relatively relaxed when confronted with an unfamiliar situation and show little indication of distress. Other children react to novel objects and situations with anxiety and try to remain close to their mother. They do not readily explore novel objects or easily interact with unfamiliar people. These actions signal behavioral inhibition. Chen et al. (1998) found cultural differences in mothers' attitudes toward such behavioral inhibition in children. Chinese mothers, who traditionally value mutual interdependence, view behavioral inhibition in toddlers as a positive trait. Behavioral inhibition is associated with maternal acceptance of the child and belief in encouraging children's achievement. That is, traditional Chinese mothers of inhibited children harbor warm and accepting attitudes toward them. In contrast, Canadian parents of European origin traditionally hold a more individualistic orientation and regard behavioral inhibition poorly; it is negatively associated with maternal acceptance and encouragement of children's achievement. Canadian mothers of inhibited children tend to be more punitive. Among Chinese families, children who display higher levels of behavioral inhibition have mothers who are less likely to believe that physical punishment is the best way to discipline them and less likely to feel angry toward them. However, maternal punishment orientation is positively correlated with behavioral inhibition in Canadian families; mothers whose children display higher levels of behavioral inhibition are more likely to believe that physical punishment is the best discipline strategy. In short, behavioral inhibition in children is associated with positive attitudes in traditional Chinese mothers and negative attitudes in Canadian mothers. In school, Chinese children who were shy achieved academically, and their teachers and peers rated them more positively; by contrast, shy Canadian children did worse in general (Chen et al., 1998). Shyness and inhibition in children are regarded by parents and teachers in Western cultures as signs of incompetence and immaturity requiring protection and intervention, whereas they are viewed by Chinese parents and teachers as desirable characteristics and are linked to optimal adjustment in Chinese children. Hence, the force of cultural representations—the implicit meanings and obvious understandings in culture, parenting, and childhood.

Two additional examples bring home this message. Maternal controlling behavior is associated with insecure attachment in U.S. American infants, whereas it is associated with secure attachment in Puerto Rican infants (Carlson & Harwood, 2003). Lansford, Chang, and Dodge (2005) conducted interviews with mother–child dyads in China, India, Italy, Kenya, the Philippines, and Thailand to examine the link between mothers' use of physical discipline and children's adjustment. Multilevel regression analyses revealed that physical discipline was less strongly associated with adverse child outcomes in conditions of greater perceived cultural normativeness. In all countries, greater reliance on physical discipline was associated with more aggression and anxiety, but countries with the lowest overall use of physical discipline showed the strongest association between mothers' use and children's behavior problems. Cultural factors (*normativeness*) moderated the association.

Different Forms–Same Functions

Different forms are sometimes end points in themselves, but they can also serve similar ultimate cultural purposes. That is, different activities also lead to culturally equivalent patterns of social adjustment. Many interesting intracultural and cross-cultural phenomena emerge when different inputs or different arrangements of particular inputs are conceived to have the same overarching cultural relevance or meaning or achieve equivalent ultimate impact on development. An assumption of different forms having the same function or meaning is that parents everywhere share and work toward some of the same ultimate goals for their children (which they do), including physical health and psychological well-being, personal happiness and social adjustment, and educational achievement and economic security, for example.

Age-graded examples of this relation may be easier to assimilate. Parents do not use the self-same behaviors to reach the same goals for their children at different points in childhood: They do not reward infants and teens the same way, for instance. Keeping an eye on a toddler in the park when the child is beginning to walk may be conceptually similar to giving a cell phone to a teen who is beginning to drive a car, in the sense of reflecting continuity in parental monitoring. To the extent that similarity exists in a given parenting domain, it resides not necessarily in the self-same parental behaviors but in the same meaning of behavior patterns or effects. On average, Japanese and U.S. American mothers are both responsive and may be equally so, and both cultural groups of mothers typically respond to the same setting events, but responsiveness of Japanese and U.S. American mothers differs in emphasis. Japanese mothers tend to use mother–infant eye-to-eye contact as a setting event to begin a face-to-face interaction or keep one going, whereas U.S. American mothers tend to use eye-to-eye contact as a setting event to introduce their infants to something in the environment outside the dyad or to label objects.

Baumrind's (1967) categories of authoritative, authoritarian, permissive, and disengaged parenting involve different emphases on style and content. She hypothesized that these practices contribute differentially to child identity formation and cognitive and moral development. Her findings showed that some parenting practices facilitate growth (e.g., social competency), whereas others do not. An authoritative style combines high levels of warmth with moderate to high levels of control. In middle-class European American children, the pattern is associated with achievement of social competence and overall better adaptation. Authoritarian parenting, by contrast, contains high levels of control, but little warmth or responsiveness to children's needs, and it is generally associated with poorer developmental outcomes. In different groups, however, different outcome patterns appear to obtain. For example, adolescents from European American and Latin American authoritative homes perform well academically and better than those coming from nonauthoritative households. However, school performance is similar for authoritatively and for nonauthoritatively reared Asian Americans (Dornbusch, Ritter, Leiderman, Roberts, & Fraleigh, 1987). Among members of one cultural group, engaging in controlling behaviors might translate into less adaptive authoritarian parenting; however, in another group in other circumstances, a style of interaction that is very directive can be adaptive rather than reflecting harsh control. Being directive may be an appropriate parental adjustment in dangerous circumstances of childhood when parents need to impress on their children and children need to know the necessity of following rules.

These interpretive matters of similarity and continuity of meaning in different forms having the same ultimate function are subtle and complex when crossing cultures. Even with generally similar ultimate goals in mind, cultures may still condition specific types of competencies that caregivers promote in children, the paths caregivers follow to instill in children the desire for achieving those goals, and the developmental timetables caregivers wish their children to meet. That is, parents might engage in different, culturally specific forms to achieve ultimately similar functions. These are questions of contextualization and specificity. Culture-specific patterns of child rearing are adapted to each specific society's settings and needs, although some different patterns may serve ultimately similar functions. Understanding relations between form and function as embedded in cultural transaction depends on context, the unit of analysis, and the level of abstraction chosen for analysis. Concretely, warmth may be expressed physically or verbally in different groups: different form–same function. In undertaking cultural analyses, one needs to remain keenly aware of these variations.

CULTURE→PARENT→CHILD

Central to the concept of culture is the expectation that different peoples possess different beliefs and behave in different ways with respect to their parenting. In turn, it is a particular and continuing task of parents (and other caregivers) to enculturate children, that is, to prepare them for socially accepted physical, economic, and psychological situations that are characteristic of the culture in which they are to survive and hopefully thrive (Benedict, 1938; LeVine, 2003). In this context, a challenge in the study of culture, parenting, child development, and transaction is how parents' child rearing is culturally shaped.

Parents' beliefs presumably relate to parents' behaviors and in turn to children's development. Expectations about developmental norms and milestones—when a child is expected to achieve a particular developmental skill—affect parents' appraisals of their child's development and of child development per se. Adult caregivers—parents, teachers, or others—normally harbor ideas about when children should be capable of certain achievements (Becker & Hall, 1989), and these ideas can themselves affect development. For example, Hopkins and Westra (1990) surveyed English, Jamaican, and Indian mothers living in the same city and found that Jamaican mothers expected their children to sit and to walk earlier, whereas Indian mothers expected their children to crawl later. In each case, children's actual attainment of developmental milestones accorded with their mother's expectations. However, parents' cognitions and practices develop in culture.

To explore this culture→parent→child chain, Cole and Tamang (1998) contrasted Chhetri-Brahmin and Tamang parents and children in Nepal. They hypothesized that cultural differences in the two groups, related to religious traditions of Hindu versus Tibetan Buddhism, would manifest themselves in parents' socialization practices and children's emotional sensitivities. Chhetri-Brahmin life emphasizes self-awareness and discipline; in contrast, Tamang culture is more egalitarian, emphasizing the importance of maintaining a calm and peaceful mind. Interviews with mothers that explored differences in their preferred socialization practices revealed that Chhetri-Brahmin mothers tended to teach their children about emotion displays, whereas Tamang mothers maintained that children learn on their own. Tamang mothers favored cajoling and comforting a young child who expressed anger toward them; Chhetri-Brahmin mothers favored reprimanding. These ethnotheories were further reflected in children's early understanding of *masking* emotions. Differences in children's choices of masking in hypothetical challenge situations were dramatic: 40% to 70% of the Chhetri-Brahmin children reported masking, whereas virtually no Tamang children did. This study instantiates a progression of ideas from cultural belief systems to parental cognitions to parents' child-rearing practices to the ideas and actions of children themselves.

CONCLUSIONS

In a transaction, at least two people interact in such a way that the intentions or actions of one reciprocally affect the intentions or actions of the other through time. The transaction principle acknowledges that an individual's characteristics shape his or her experiences, and reciprocally, experiences shape the characteristics of the individual through time. By virtue of their unique characteristics and propensities children actively contribute to producing their own development. For their part, parents influence their own and their children's development through many different means. Parenting is best conceptualized as multidimensional, modular, and specific, and having meaning that reflects its embedding in a specific culture (Bornstein, 2006).

The specificity principle states that specific experiences specific parents provide specific children at specific times exert effects in specific ways over specific aspects of child development (Bornstein, 2002, 2006). The specificity principle accords with a situational contextual—culturally informed and sensitive—view of parenting. Historically, parents have been cast as the primary agents who set the agenda for what children learn and who administer the rewards and punishments that strengthen desired characteristics and weaken undesired ones in children. So, characteristics of parents are influential. Many theoretical accounts for how socialization takes place have been proposed. For most theories, it is primarily through parental action that the adult culture is passed down to each new generation of children. In a larger sense, parenting cognitions and parenting practices contribute to the continuity of culture by helping to define culture and the transmission of culture across generations. Cultural variations in child rearing exert significant and differential influences over physical, mental, emotional, and social development of children, just as cultural variation dictates the language children eventually speak.

Child development is influenced by multiple factors. Within any culture, children are shaped by the physical and social settings within which they live, culturally regulated customs and child-rearing practices, and culturally based belief systems. The psychological meaning attributed to any given behavior is, in large measure, a function of the ecological niche within which it is produced. If a given behavior is viewed as culturally desirable or acceptable, then parents (and significant others) will encourage its development; if the behavior is perceived as culturally maladaptive or abnormal, then parents (and significant others) will discourage its development. There are multiple and distinctive pathways for socializing children to become competent adults, and optimal development is largely defined by the cultural meaning system (Bornstein, 1995).

Assumptions about the specificity and generality of human development may be most advantageously tested within the context of intracultural and cross-cultural transactional developmental research because parents and children do not develop in a vacuum, but their relationships grow in the media of one or more environments, physical and cultural. The range and diversity of normal human child environments are just about as vast as is imaginable. It is only by looking at multiple beliefs and behaviors in multiple physical and cultural contexts that we are afforded the possibility to appreciate developmental transactions.

REFERENCES

Amato, P. R., & Booth, A. (2001). The legacy of parents' marital discord: Consequences for children's marital quality. *Journal of Personality and Social Psychology*, *81*, 627–638.

Anderson, K. E., Lytton, H., & Romney, D. M. (1986). Mothers' interactions with normal and conduct-disordered boys: Who affects whom? *Developmental Psychology, 22,* 604–609.

Bates, J. E., Pettit, G. S., & Dodge, K. A. (1995) Family and child factors in stability and change in children's aggressiveness in elementary school. In J. McCord, (Ed.), *Coercion and punishment in long-term perspectives* (pp. 124–138). New York: Cambridge University Press.

Baumrind, D. (1967). Socialization practices associated with dimensions of competence in preschool boys and girls. *Child Development, 38,* 291–327

Becker, J. A., & Hall, M. S. (1989). Adult beliefs about pragmatic development. *Journal of Applied Developmental Psychology, 10,* 1–17.

Bell, R. Q. (1968). A reinterpretation of the direction of effects in studies of socialization. *Psychological Review, 75,* 81–95.

Bell, R. Q., & Harper, L. (1977). *Child effects on adults.* Hillsdale, NJ: Erlbaum.

Belsky, J., Hsieh, K. H., & Crnic, K. (1998). Mothering, fathering, and infant negativity as antecedents of boys' externalizing problems and inhibition at age 3 years: Differential susceptibility to rearing experience? *Development and Psychopathology, 10,* 301–319.

Benedict, R. (1938), Continuities and discontinuities in cultural conditioning. *Psychiatry, 1,* 161–167.

Biringen, Z., Emde, R. N., Campos, J. J., & Appelbaum, M. I. (1995) Affective reorganization in the infant, the mother, and the dyad: The role of upright locomotion and its timing. *Child Development, 66,* 499–514.

Bornstein, M. H. (1995). Form and function: Implications for studies of culture and human development. *Culture & Psychology, 1,* 123–137.

Bornstein, M. H. (2002). Parenting infants. In M. H. Bornstein (Ed.), *Handbook of parenting: Vol. 1. Children and parenting* (2nd ed., pp. 3–43). Mahwah, NJ: Erlbaum.

Bornstein, M. H. (2006). Parenting science and practice. In I. E. Sigel & K. A. Renninger (Vol. Eds.) & W. Damon & R. M. Lerner (Series Eds.), *Handbook of child psychology: Vol. 4. Child psychology and practice* (6th ed., pp. 893–949). New York: Wiley.

Bornstein, M. H., Hahn, C.-S., Bell, C., Haynes, O. M., Slater, A., Golding, J., et al. (2006). Stability in cognition from early infancy: A developmental cascade. *Psychological Science, 17,* 151–158.

Bornstein, M. H., Haynes, O. M., Azuma, H., Galperín, C., Maital, S., Ogino, M., et al. (1998). A cross-national study of self-evaluations and attributions in parenting: Argentina, Belgium, France, Israel, Italy, Japan, and the United States. *Developmental Psychology, 34,* 662–676.

Bornstein, M. H., Tal, J., Rahn, C., Galperín, C. Z., Pêcheux, M.-G., Lamour, M., et al. (1992). Functional analysis of the contents of maternal speech to infants of 5 and 13 months in four cultures: Argentina, France, Japan, and the United States. *Developmental Psychology, 28,* 593–603.

Carlson, V. J., & Harwood, R. L. (2003). Alternate pathways to competence: Culture and early attachment relationships. In S. M. Johnson & V. E. Whiffen (Eds.), *Attachment processes in couple and family therapy* (pp. 85–99). New York: Guilford Press.

Chase-Lansdale, P. L., Cherlin, A. J., & Kiernan, K. K. (1995). The long-term effects of parental divorce on the mental health of young adults: A developmental perspective. *Child Development, 66*, 1614–1634.

Chen, X., Hastings, P., Rubin, K. H., Chen, H., Cen, G., & Stewart, S. L. (1998). Child-rearing attitudes and behavioral inhibition in Chinese and Canadian toddlers: A cross-cultural study. *Developmental Psychology, 34*, 677–686.

Cole, P. M., & Tamang, B. L. (1998). Nepali children's ideas about emotional displays in hypothetical challenges. *Developmental Psychology, 34*, 640–646.

Dallaire, D. H., & Weinraub, M. (2005). The stability of parenting behaviors over the first 6 years of life. *Early Childhood Research Quarterly, 20*, 201–219.

Deci, E. L., Ryan, R. M., & Koestner, R. (1999). A meta-analytic review of experiments examining the effects of extrinsic rewards on intrinsic motivation. *Psychological Bulletin, 125*, 627–668.

DeVries, M. W. (1984). Temperament and infant mortality among the Masai of East Africa. *American Journal of Psychiatry, 141*, 1189–1194.

Dornbusch, S. M., Ritter, P. L., Leiderman, P. H., Roberts, D. F., & Fraleigh, M. J. (1987). The relation of parenting style to adolescent school performance. *Child Development, 58*, 1244–1257.

Farrington, D. P. (1994). *Psychological explanations of crime*. Brookfield, VT: Dartmouth Publishing.

Fraiberg, S., Adelson, E., & Shapiro, V. (2003). Ghosts in the nursery: A psychoanalytic approach to the problems of impaired infant–mother relationships. In J. Raphael-Leff (Ed.), *Parent–infant psychodynamics: Wild things, mirrors and ghosts* (pp. 87–117). London: Whurr.

Gandour, M. J. (1989). Activity level as a dimension of temperament in toddlers: Its relevance for the organismic specificity hypothesis. *Child Development, 60*, 1092–1098.

Herz, L., & Gullone, E. (1999). The relationship between self-esteem and parenting style: A cross-cultural comparison of Australian and Vietnamese Australian adolescents. *Journal of Cross-Cultural Psychology, 30*, 742–761.

Holden, G. W., & Miller, P. C. (1999). Enduring and different: A meta-analysis of the similarity in parents' child rearing. *Psychological Bulletin, 125*, 223–254.

Hopkins, B., & Westra, T. (1990). Motor development, maternal expectations, and the role of handling. *Infant Behavior & Development, 13*, 117–122.

Kotchabhakdi, N. J., Winichagoon, P., Smitasiri, S., Dhanamitta, S., & Valyasevi, A. (1987). The integration of psychosocial components in nutrition education in northeastern Thai villages. *Asia-Pacific Journal of Public Health, 1*, 16–25.

Kuczynski, L., Kochanska, G., Radke-Yarrow, M., & Girnius-Brown, O. (1987). A developmental interpretation of young children's noncompliance. *Developmental Psychology, 23*, 799–806.

Lansford, J. E., Chang, L., & Dodge, K. A. (2005). Physical discipline and children's adjustment: Cultural normativeness as a moderator. *Child Development, 76*, 1234–1246.

LeVine, R. A. (2003). *Childhood socialization: Comparative studies of parenting, learning and educational change*. Hong Kong, China: University of Hong Kong.

Lindahl, K. M., & Malik, N. M. (1999). Marital conflict, family processes, and boys' externalizing behavior in Hispanic American and European American families. *Journal of Clinical Child Psychology, 28*, 12–24.

Maccoby, E. E., & Martin, J. A. (1983). Socialization in the context of the family: Parent–child interaction. In E. M. Hetherington (Vol. Ed.) & P. H. Mussen (Series Ed.), *Handbook of child psychology: Vol. 4. Socialization, personality, and social development* (pp. 1–101). New York: Wiley.

Mondschein, E. R., Adolph, K. E., & Tamis-LeMonda, C. S. (2000). Gender bias in mothers' expectations about infant crawling. *Journal of Experimental Child Psychology, 77*, 304–316.

Newcomb, M. D., & Locke, T. F. (2001). Intergenerational cycle of maltreatment: A popular concept obscured by methodological limitations. *Child Abuse & Neglect, 25*, 1219–1240.

Noller, P., Seth-Smith, M., Bouma, R., & Schweitzer, R. (1992). Parent and adolescent perceptions of family functioning: A comparison of clinic and non-clinic families. *Journal of Adolescence, 15*, 101–114.

Ochs, E., (1987). The impact of stratification and socialization on men's and women's speech in Western Samoa. In S. U. Philips, S. Steele, & C. Tanz (Eds.), *Language, gender, and sex in comparative perspective* (pp. 50–70). New York: Cambridge University Press,

Pachter, L. M., & Dworkin, P. H. (1997). Maternal expectations about normal child development in 4 cultural groups. *Archives of Pediatrics and Adolescent Medicine, 151*, 1144–1150.

Quinn, N., & Holland, D. (1987). Culture and cognition. In D. Holland & N. Quinn (Eds.), *Cultural models in language and thought* (pp. 1–40). New York: Cambridge University Press.

Rohner, R. P., & Pettengill, S. M. (Eds.). (1985). Family development [Special issue]. *Child Development, 56*(2).

Rudy, D., & Grusec, J. E. (2006). Authoritarian parenting in individualist and collectivist groups: Associations with maternal emotion and cognition and children's self-esteem. *Journal of Family Psychology, 20*, 68–78.

Ruoppila, I. (1991). The significance of grandparents for the formation of family relations. In P. K. Smith (Ed.), *The psychology of grandparenthood: An international perspective*. (pp. 123–139). Florence, KY, Taylor & Frances and Routledge.

Scarr, S., & Kidd, K. K. (1983). Developmental behavior genetics. In M. M. Haith & J. J. Campos (Vol. Eds.) & P. H. Mussen (Series Ed.), *Handbook of child psychology: Vol. 2. Infancy and developmental psychobiology* (pp. 345–433). New York: Wiley.

Scarr, S., & McCartney, K. (1983). How people make their own environments: A theory of genotype (leading to) environment effects. *Child Development, 54*, 424–435.

Schank, R., & Abelson, R. (1977). *Scripts, plans, goals and understanding: An inquiry into human knowledge structures*. Hillsdale, NJ: Erlbaum.

Smith, P. K., & Drew, L. M. (2002). Grandparenthood. In M. H. Bornstein (Ed.), *Handbook of parenting: Vol. 3. Status and social conditions of parenting* (2nd ed., pp. 141–172). Mahwah, NJ: Erlbaum.

Stevenson, H. W., Lee, S., & Chen, C. (1990). Contexts of achievement: A study of American, Chinese, and Japanese children. *Monographs of the Society for Research in Child Development, 55*(1–2), 1–123.

Stith, S. M., Rosen, K. H., & Middleton, K. A. (2000). The intergenerational transmission of spouse abuse: A meta-analysis. *Journal of Marriage & the Family, 62*, 640–654.

Tamis-LeMonda, C. S., & McFadden, D. (in press). The United States. In M. H. Bornstein (Ed.), *The handbook of cross-cultural developmental science: Vol. 2. Human development in different places on earth.* New York: Psychology Press.

Thomas, A., Chess, S., & Birch, H. G. (1968). *Temperament and behavior disorders in children*. New York: New York University Press.

van IJzendoorn, M. H. (1992). Intergenerational transmission of parenting: A review of studies in nonclinical populations. *Developmental Review, 12*, 76–99.

Vermulst, A. A., de Brock, A. J. L. L., & van Zutphen, R. A. H. (1991). Transmission of parenting across generations. In P. K. Smith (Ed.), *The psychology of grandparenthood: An international perspective* (pp. 100–122). Florence, KY, Taylor & Frances and Routledge.

Vygotsky, L. (1978). *Mind in society*. Cambridge, MA: Harvard University Press.

9

SOCIAL AND CULTURAL TRANSACTIONS IN COGNITIVE DEVELOPMENT: A CROSS-GENERATIONAL VIEW

MARY GAUVAIN

The transactional model has been used to describe psychological development in many areas, including caregiver–child interactions, child socialization, and maladjustment. The chapters in this volume recount the wide-ranging contribution of this approach to theory and research. The present chapter follows in this tradition; however, it adopts a longer range view. It discusses how social psychological transactions, especially those important to cognitive development, contribute to the cultural context of human development. The chapter is particularly concerned with the cultural context of cognitive development as it transcends generations.

The study of cognitive development is in many ways the study of culture. Human beings possess the ability to participate in historically formulated courses of intelligent action, the products of culture. Culture provides the experiences and artifacts that mediate mental activity, and over the course of development these mediational means affect how children understand and interact with the world (Cole, 1996). Children learn about culture in many ways: with the help of more experienced members, by participating in valued practices and institutions, and by using representational and material tools (Gauvain, 1995). In these situations of learning, ways of thinking and solving problems that have been successful in the past and remain of value are transmitted to children. Thus, as children come to understand and adopt these practices and tools, culture is embodied in the behavior of new cultural members and sustained across generations. Or so it would seem.

Cultures cannot be sustained solely through the transmission of cultural practices and tools. Through both circumstance and design, cultures change, and as a result, transmission across generations of prior ways of understanding and acting on the world is inadequate. For a culture to be sustained, it also needs to change. Yet how do cultural changes come about? And do these changes relate to, perhaps rely on, processes of psychological development?

In this chapter, I take the position that cognitive development is an important force in cultural change. I use the transactional model (Sameroff, 1989) to describe how the social and psychological processes through which children learn about and participate in culture contribute to cultural stability and change across generations.

CULTURAL MAINTENANCE AND CULTURAL CHANGE

Human beings live in culture; it is the species-specific habitat—the setting in which human beings naturally live and grow. The need for children to be part of culture and for culture to maintain and build on its accomplishments across generations vis-à-vis its members are complementary and generative forces. There are many definitions of culture. Some emphasize community structure; others favor historical description; and others concentrate on psychological features, such as meaning and belief systems. Despite differences, all definitions include implicitly or explicitly that cultures are sustained social groups that depend on the transmission across generations of values and practices of the group. In cultural transmission, established ways of organizing, interpreting, and carrying out human activity are transferred to new generations. Yet right alongside efforts to maintain values and practices, there are pressures for cultural change.

In all cultures, the natural course of human life dictates change as old members pass on and new members assume roles and responsibilities. Cultural change also results from fluctuations in the environment, such as alterations in the ecosystem or natural resources. It stems from outside social forces, including cultural contact (Gauvain, Altman, & Fahim, 1983), economic reorganization (Greenfield, 2004), migration (Berry, 2007), military conquest (Erikson, 1963), modernization (Inkeles & Smith, 1974), and urbanization (Kagitçibasi, 2007). Change also comes from within communities. In some cases, it represents an intentional effort to redirect the culture, as in educational interventions such as Head Start in the United States (Zigler & Hall, 2000) and the Turkish Early Enrichment Project (Kagitçibasi, 2007). In other cases, change reflects the rejection of conventional practices, the adoption of new practices, or some type of innovation. Sometimes change that emerges from within culture is due to the actions of an individual, for example, an inventor. In other instances, it comes about through interpersonal processes, for example, when people collaborate, or remote social experiences, for example, when someone builds on the work of others, as captured in Isaac Newton's oft-quoted statement, "If I have seen further it is by standing on the shoulders of giants."

In this chapter I do not examine cultural changes that are externally driven or introduced by society at large. I concentrate on changes that come from within as community members interact with one another and with the signs and tools of the culture in the context of child socialization.

Child Socialization, Cognitive Development, and Cultural Change

During socialization, children are introduced to cultural ways of acting and thinking. They adopt some of these conventions, adapt others, discard others, and introduce new and innovative forms—all with the aim of meeting the local circumstances of growth. In most cases, these behaviors do not lead to cultural change, though they may have profound effects on individual development, a topic taken up by several chapters in this volume. However, sometimes these behaviors ignite a more collective effort, and as a result they may lead to change that extends beyond individuals and even transcends generation.

Early research on socialization concentrated on the impact of adults, especially parents, on development (Grusec & Hastings, 2007). Contemporary theories of socialization hold that children are active agents who direct the course of their own development (Kuczynski, 2003). Agents of socialization, such as parents, teachers, and peers, do not mold the child; rather, children intentionally try to understand and explore the world. And just as other people try to shape child behavior, children also modify the behaviors of others. This bidirectional process is evident in research by Chavajay and Rogoff (1999), who compared the attention-related behaviors of toddlers in a Mayan community in Guatemala with those in a middle-class community in the United States. They found that Mayan children often attended to several events simultaneously, whereas U.S. children usually attended to one event or alternated back and forth between two events. The researchers tied these patterns to differing socialization practices, especially schooling, and argued that emphasis in the United States on attending to one thing at a time helps prepare children for this institutional setting.

These results support the view that children actively adopt the cognitive practices of their culture and that through participation in these practices they maintain the values and behaviors of the group. However, these results do not provide information about the related question: How do children contribute to cultural change? A theoretical framework for answering this question was provided by Vygotsky in his writings on the role of cultural history in psychological development (Wertsch, 1998). This framework takes form in the sociocultural approach to cognitive development, which contends that over development, children engage in and, thereby, learn about culture in active and transformative ways. In this chapter, I take the position that this process is bidirectional, that is, culture transforms cognitive development, and over time, cognitive development transforms culture. Thus, from a cross-generational perspective, cognitive development plays a dual role: It helps maintain culture, and it helps change culture.

Is Childhood a Wellspring of Cultural Change?

The observation that cultures need to change to be sustained begs the following question: What do cultures have in continual supply that can motivate change and help steer change across generations? Relying on external forces or environmental factors as the primary sources of change is insufficient and runs counter to current views about the importance of agency in human psychology and its development. Something inherent to culture is also needed.

In this chapter, I argue that characteristics of childhood and the social processes in which children engage as they learn about and participate in culture operate as wellsprings of cultural change. In *Merriam-Webster's Dictionary* (2003), a *wellspring* is defined as a source of continual supply. Children's inexperience with the world, the limited nature of their cognition and responsibilities, their strivings for autonomy, their affinity toward peers, and their propensity toward play all provide availability or openness to experiences that are markedly different from the experiences adults have. These characteristics of childhood offer unique opportunities for cultural change as seen in the rapid proliferation of cell phones and iPods among youth and their associated changes to communication and learning.

Bjorklund (2007) struck a similar chord in his analysis of the role that immaturity plays in human development: "Youth is not wasted on the young. Infants and children make good use of their unique features (particularly psychological ones) both to survive the niche of childhood and to prepare themselves for life as adults" (p. 2). And decades ago, the anthropologist Barnett (1953) made a like argument:

> Many individuals are prepared to accept new ideas because they have not dedicated themselves irretrievably to a custom or to an ideal of their society. . . . The greatest number of individuals in this category are children . . . young people in any society are inevitably less securely bonded to their culture than are their seniors, if for no other reason than that they have had less time to understand and be conditioned by it. (p. 385)

Barnett backed up this claim with anthropological research and with historical anecdotes like the following. The 17th century English physician William Harvey, viewed by many as laying the foundation of modern medicine, commented that no one over 40 years of age accepted his discovery of blood circulation when it was announced. Similarly, Sameroff (personal communication, July 2007) reported that none of his colleagues read or were conversant with the transactional model until they were forced to keep up with their students.

In short, cultural change cannot be left entirely to external forces. It needs to be self-generated or come from within the culture in some way. Psychological processes, especially processes that play major roles in child development, may serve as engines of such change. More specifically, certain social and psychological processes of cognitive development may create opportunities for cultures to change, even radically so, in the brief time span of a generation.

The idea that the impetus and opportunities for cultural change are built into human psychological development does not mean that the cross-generational changes that emerge from these processes are planned or have lasting benefit. The processes that support and guide change occur in local situations—in the ways that people interact, the areas of mental functioning that are stressed, and the practices and tools people use. These processes draw on and are largely directed toward the skills and knowledge that are needed in that setting at that time. Though a change may portend the foreseeable future and even represent in some way the desired future of the culture, the ramifications of any given change are not easy to predict and in many instances they may arise gradually and even seep into the culture in unexpected ways. Who foresaw that freeway drivers, especially inexperienced teen drivers, would not only dial portable phones but watch movies on them while they were behind the wheel?

To frame how time intersects with human development in cultural context, Cole (1996) drew on Gottlieb's (1992) probabilistic view of development. In Cole's view, agents of socialization, like parents, represent the child's future in present experience. They do so in a variety of ways, by referring to future states in conversations with children, for example, "when you're a mommy"; in training efforts, for example, teaching children about social relationships and manners so they will behave properly when they interact with others (Lancy, 1996); and in the context of regular practices, for example, preparing preschoolers for the classroom by engaging them in routines of discourse that mimic school lessons (Rogoff, 2003). In these exchanges, more experienced cultural members assume cultural stability to "project a probable future for the child" (Cole, p. 184), which then informs the actual experiences children have as they grow up. In this description, Cole explained how socialization draws on the culture's past to imagine and help create the future. It refers to aspects of culture that are maintained or that people try to maintain across generations.

The present discussion extends these ideas by suggesting that cultural change, at least some types of change, is a normative feature of development. In particular, I argue that social and psychological processes that contribute to cognitive development include ways both of maintaining and of changing culture. The transactional model is important to this discussion because it can account simultaneously for transformations in psychological experience and in culture.

COGNITIVE DEVELOPMENT AND CULTURAL CHANGE

How do culture and cognitive development inform and transform each other across generations? The social and psychological experiences through which children learn about and participate in culture provide the contexts in which cultural practices and tools are scrutinized, adopted, modified, and sometimes abandoned by children. The transactional model, with its emphasis on change as a result of the dynamic coordination of elements of a complex system, is ideal for describing this process (Sameroff, 1989). In the transactional model, both the child and the environment, which for present purposes is defined as experienced cultural members and representations of culture in myriad forms, are active elements of a system that simultaneously tends toward stability (maintenance) and creates avenues of adaptation (change). Thus, change is inherent to the system; it emerges from the dynamic exchanges that occur across levels of the system (Altman & Rogoff, 1987).

Figure 9.1 depicts the transaction of psychological development and culture across generations (represented as 1 to 4 in the boxes). Consistent with views put forth by Vygotsky (1987), the figure shows an indirect or medi-

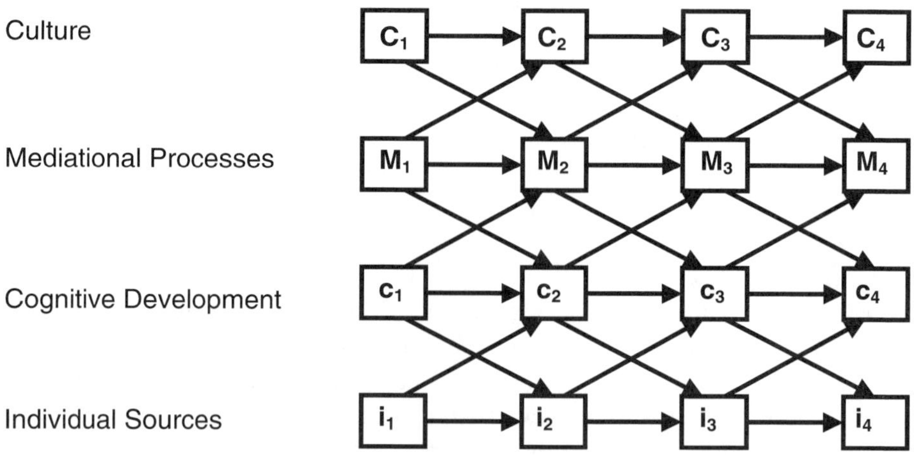

Figure 9.1. Cross-generational transactional model of culture and cognitive development. The numbers 1 through 4 refer to generations.

ated relation between culture (represented as C) and cognitive development (represented as c). Change in culture and in cognitive development across generations emerges from or is mediated by human transactions and representational forms (represented as M). In other words, change, at least the type of change discussed in this chapter, occurs at both the individual and group levels as children engage in social interactions and other types of activities that involve the practices, values, and tools of the culture.

Characteristics associated with, but not necessarily inherent to, the child (represented as i), including biological and emotional factors, also contribute. Some of these characteristics transcend generations, like genetics, and others exert primary force within a generation, like emotional regulation and compliance. Although represented in the model, these characteristics are not discussed further in this chapter. Bornstein (see chap. 8, this volume) discusses many examples of interactions between individual qualities and socialization that vary across cultures.

The aim of this chapter is to use the transactional model to describe how cognitive development contributes to cultural stability and change across generations. The next section describes the sociocultural approach to cognitive development (Cole, 1996; Rogoff, 2003), which is foundational to the argument I present. Some relevant ideas from evolutionary developmental psychology are also included. This is followed by research on social processes of cognitive development—processes that connect children's thinking to culture and, as a result, operate as psychological mechanisms of cultural maintenance and change.

The Social and Cultural Nature of Cognitive Development

Much of the research on cognitive development focuses on what children know and how they process information on their own. An important related question pertains to the social and cultural basis of this knowledge, both in terms of content and the means by which it develops.

To reach adult levels of competence, children have to learn a lot about their culture. To this end, cultural members rely on many social learning processes, including observation, imitation, attention regulation, demonstration, instruction, rehearsal, shaping, scaffolding, guided participation, and trial and error. In every culture there are also established settings in which learning occurs, including institutions such as school, and other organized settings, such as apprenticeships, routines, and rituals (Serpell & Hatano, 1997). These settings contain actions that are shared by and have meaning to group members (Greenfield & Lave, 1982), are repeated and function as regular parts of the culture (Goodnow, Miller, & Kessel, 1995), and include normative expectations that extend beyond the immediate situation and guide subsequent behavior (Valsiner & Lawrence, 1997). Repeated experiences facilitate cultural learning, and over time it is these regular transactions that define the nature and course of human development in a particular cultural context (White, 2000).

Although the sociocultural approach views social and cognitive development as mutually constituted, it does not claim that biology and other internal contributions, such as emotional processes, are unimportant. These contributions help set the boundaries and pace of development and activity at any given point in growth. Also, the sociocultural perspective is not alone in emphasizing the interplay of internal capabilities and external factors; ecological systems theory also assumes this stance (Bronfenbrenner & Morris, 2006). However, a sociocultural approach is unique in its attention to specific aspects of the social and cultural context that form the basis of intellectual development.

In the sociocultural approach, the context of cognitive development is defined in broad terms and includes social interaction; the signs and tools that people use to support and extend thinking; and cultural values instantiated in routines, rituals, and customs. Cognitive development is not the direct consequence of sociocultural experience. Rather, the transactions that compose human social experience generate this development. As in the transactional model, cognitive development is a dynamic process, the outcome of which cannot be known beforehand or by examining the individual or social context alone. Also, the specific social processes and cultural goals that guide cognitive development at any point in time are flexible and respond to local conditions. As a result, it is expected that the content, procedures, and participants vary across cultural settings and reflect the time, place, and needs of individuals and the community.

As children learn about the world through social experiences and through activities mediated by cultural signs and tools, their thinking comes to incorporate the understanding and practices of the culture. This culturally mediated understanding then joins with the direct and natural experiences children have to shape cognitive development. Cultural history as it exists in the signs and tools that mediate human mental functioning is especially important to this process. Cultural signs, such as language and the number system, and tools, such as navigational systems and technology, affect how people interact with the world. These signs and tools, which are transmitted across generations, are not merely external stimuli to which children learn to respond; they carry meaning, and it is the meaning that is learned and understood by children. It is significant that the signs and tools that are part of cognitive development are not arbitrary but reflect an organized and historical system, namely culture, and thereby contain psychological connections to other societal members and ancestors. Participation in this meaning system enables children to interact with others as well as interpret and act on the world in ways that make sense to other people in the culture. It may also facilitate the development of emotional connections and allegiances to the culture and its history.

The gradual incorporation of signs and tools into mental functioning are evident in a child's development. As a child's capabilities, interests, and needs change, the way that signs and tools are used to mediate action also changes. Consider the many transformations in the use of writing and the number system over childhood; what begins as a set of shapes to recognize, copy, and draw changes into a means of organizing, manipulating, conveying, and retaining thoughts. The use of signs and tools also undergoes change at the cultural level, for example, when the group confronts new life circumstances. Inkeles and Smith (1974) found that the presence and use of clocks transformed when a nation's economic base shifted from agricultural to industrial.

Although human beings are active agents, there are constraints at all levels—the child, the culture, and the mediational means—and these constraints influence the transactions that support cognitive development and cultural change. As a result, these transactions are not chaotic, random, or aimless. They are regulated by many internal and external factors, and certain forms of experience or input have priority. Innate human capabilities, including the ability to use language, communicate, and engage in social relationships (LeVine et al., 1994), along with features of the cognition system that regulate the information that human beings attend to and how it is processed (Elman et al., 1996), help determine the nature and effectiveness of these transactions. Individual factors, including developmental status, learning history, and cognitive abilities, are also important. Cultural values, practices, symbols and artifacts, communicative conventions, social relations and behaviors, and the content and purpose of a learning activity also matter. The ideas that a culture has about knowledge, what is important to know, and how knowledge develops are also significant (Goodnow, 1990). These various constraints reflect human evolutionary history.

Human Evolution in Relation to the Social and Cultural Basis of Cognitive Development

Although phylogenetic, historical, and ontogenetic changes occur on vastly different time scales, an evolutionary perspective is consistent with the view of cognitive development as a social and cultural process. Human beings have evolved capabilities to learn a vast array of complex skills and information. As stated previously, these capabilities are biased toward certain types of information and patterns and they include the ability to devise and use symbolic and material tools to support and extend intelligent action. These abilities help human beings solve problems critical to survival, such as finding a mate, hunting for food, and rearing the young (Barkow, Cosmides, & Tooby, 1992).

For evolutionary psychologists, human evolution is largely about cognition. However, the evolution of human intelligence is inseparable from two other species characteristics: the social nature of human beings and the fact that high levels of cognitive functioning come about in the lifetime of an individual. During human evolution, cognitive abilities emerged that enabled complex social forms, such as recognizing and communicating with group members, cooperating for resources, and understanding the intentions of others (Tomasello, Kruger, & Ratner, 1993). The social bias present in early perceptual development reflects these abilities and helps children learn about the world from other people from the very beginning of life (Gauvain, 2001).

The means by which human intelligence develops in an individual's lifetime were also crafted by evolution. Human beings are born with an immature brain; they have a protracted period of dependence; and they are capable of forming affectional ties to others (Bjorklund & Pellegrini, 2002; Hinde, 1987). These characteristics provide children with substantial contact with more mature cultural members at precisely the time when much learning occurs. Caregivers know much of what children need to learn to survive in their habitat and these contacts involve strong emotional ties, which enhance the potential for learning by increasing arousal and regulating attention. Thus, as the child develops, the social context of development results in psychological functioning specialized to the unique circumstances of growth. This transformation, which Vygotsky (1981) called "the general law of cultural development," is the gradual shift over development from the interpsychological, or understanding between people, to the intrapsychological, or understanding within the child.

In short, an evolutionary view provides a historical and biological base for the link between cognitive development and social and cultural processes. In the next section, I rely on assumptions put forward in the sociocultural approach and the transactional model to describe how cultural maintenance and change may be realized through psychological development.

PROCESSES OF COGNITIVE DEVELOPMENT IN SOCIAL CONTEXT RELATED TO CULTURAL MAINTENANCE AND CHANGE

Cognitive development is the process of socialization to the intellectual life of the community in which children are expected to become mature and competent members (Gauvain & Perez, 2007). The social world contributes to cognitive development in three ways: It determines what children think about; it influences how children think; and it provides the social processes through which children learn about the world and develop cognitive skills The social processes cover a wide spectrum of children's experiences, including interpersonal interactions and more distal social processes like participation in culturally organized practices and the use of cultural tools and symbol systems. The range of these experiences underscores the variety and richness of human life as expressed in the social experiences that support the development of thinking. They also point to the intricate link between culture and cognitive development.

Contemporary research describes many social processes that contribute to cognitive development. Some are interpersonal and entail explicit efforts to convey information or instruct, such as apprenticeship (Lave, 1988), scaffolding (Wood & Middleton, 1975), and interaction in the zone of proximal development (Vygotsky, 1987). Other interpersonal processes involve child participation in intelligent action, though these processes may not have explicit instructional goals, for example, conversations about the past and as events unfold (Nelson, 2007), guided discovery (D'Andrade, 1981), guided participation (Rogoff, 1990), shared intentions (Tomasello, Carpenter, Call, Behne, & Moll, 2005), joint attention (Bakeman & Adamson, 1984), listening in (Rogoff, 2003), participation in routines and rituals (Fiese, 2006), and social referencing (Campos & Stenberg, 1981). Still other processes are social in the sense that they involve watching other people, but they do not include interpersonal exchange, for example, legitimate peripheral participation (Lave & Wenger, 1991) and social observational learning (Bandura, 1986). Finally, some processes may not involve other people at all, but they are social in the sense that they involve representational forms derived from culture, for example, conventions of language (Schiefflin & Ochs, 1986); established problem-solving routines, for example, maps, plans, algorithms (Goody, 1977); literacy and other symbolic forms (Olson, 1994); and technology and its artifacts (Turkle, 1984).

Despite differences in these processes, they are similar in that each embodies to some degree the values and practices of the culture. As a result, over the course of development as children engage in these processes they appropriate the practices and knowledge of their culture. When these social processes draw on established practices and understanding, the purpose is to maintain them in the next generation. This may especially be the case in social interactions involving partners of unequal status such as parents and children, teachers and students, and masters and apprentices. More experienced cultural members are often charged with this goal and if the child comes to understand and adopt the cultural information or skill conveyed, these interactions succeed in meeting this goal.

When this outcome occurs, it is likely that certain social and psychological conditions were met. For interpersonal processes it may be crucial for partners to establish intersubjectivity (Rommetveit, 1985) and for children to comply with requests from more experienced partners (Maccoby, 1992). For participatory learning processes, it may be essential that the learner understand the intentions of other people. In observational learning it is necessary that a behavior is accessible; behaviors that rely on mental processes, such as decision making, are not amenable to this type of social learning. Representational and material tools often have design features that direct the process and outcome of learning (White, 2000). When this information is not transparent, more explicit information may be provided, for example, directions or a user's guide. If a child uses tools in ways not intended by the culture, a more experienced member may step in and realign the child's activity along culturally desired lines—saying, for example, "Let me show you how to use that" (Gauvain, de la Ossa, & Hurtado, 2001).

Peer Interactions

So far, my examination of these processes provides information about how culture is maintained through social and psychological means that support and direct cognitive development. It does not provide much information about the social and psychological processes behind cultural change. The asymmetrical relations between children and more experienced cultural members, an arrangement that has been the focus of much of the research based on the sociocultural approach, may generate learning opportunities that are primarily relevant to cultural transmission. Interaction with peers may be an entirely different story. Peer interaction has the potential to be more open and egalitarian than interactions involving partners of unequal status. As a result, peer interaction may be particularly relevant to cultural change.

Of course, like adults, children have developed the skills and values of their culture, and research has shown that this understanding can sometimes guide peer interaction; for example, some preschoolers assume the role of "gender enforcer," making sure that cultural rules pertaining to gender are followed when girls and boys play together (Fagot, 1977). However, peer interactions can also lead to social processes that challenge or undermine conventional forms and, thereby, function as inroads of cultural change. These challenges may occur inadvertently, perhaps as a result of characteristics that children bring to the interactions. Children's inexperience with the world, including limited understanding of cultural conventions, developing yet not fully formed (or socialized) regulatory skills, a need for independence, and interest in play, may provide an openness to experience that invites change. When children interact they may also intentionally promote change, for example, by modifying or rejecting conventional behaviors or by introducing new forms. Research on teen chat rooms on the Internet suggests that adult-monitored sites involve different types of participants and generate far different exchanges and potential learning opportunities for youth than unmonitored sites (Subrahmanyam, Smahel, & Greenfield, 2006).

What peer processes may challenge the status quo and have the potential to promote different ways of thinking and acting than cultural conventions prescribe? Research suggests that some peer processes may be especially important in this regard, such as collaboration and competition (Tudge, 1992), conformity/nonconformity (Serpell, 1993), negotiation (Serpell, 1995), and risk taking (Lightfoot, 1997). All have been linked to novel cognitive contributions when children work together. Novelty may be especially likely in peer situations that involve material artifacts. Although tools tend to endorse cultural maintenance, in learning to use a tool behaviors can emerge that modify the tool or create a new "off-shoot" tool (Tomasello et al., 1993). Such transformations can arise spontaneously from trial and error or from play, or they may be sought out through strategic exploration or manipulation of the tool.

Of course, an important question is whether such inventive forays are more likely when peers work together as compared with adult–child interaction or solitary child activity, as proposed here. Some research suggests that they may be. For example, research on the use of technology in literacy development found that peer interaction, even when simulated, led to active language construction by young children (Cassell, 2004). The rapid history and development of the Web-based communication Facebook may serve as another such illustration. Or consider that in 1976 Steve Jobs, age 21, and Steve Wozniak, age 26, combined their skills to form Apple Computer and popularize the personal computer, a collaboration that spawned a cultural transformation (Carlton, 1997). Peer processes that promote innovative cultural forms may be pronounced at certain points in development, perhaps especially in adolescence when peer influence, autonomy concerns, and cognitive capabilities are at a unique juncture (Steinberg, 2007).

Drawing attention to specific age groups does not discount the insight and wisdom that may come with longevity. But it does underscore the point that viewing child and adolescent development as only preparatory states for adulthood may overlook the unique and vital contributions these age groups make to human development and to culture (Bjorklund, 2007).

Adult–Child Interaction

Even though this discussion has emphasized the contribution of peer processes to cultural change, peer interaction is not the only social source of such change. Sometimes parents strive, as is the case in many immigrant families, to make a life for their children that is very different from the one the parents themselves experienced. In addition, when adults and children work together, differences in expertise may generate processes that override age-related status and affect change. That many youth know more about computers and other forms of technology than their parents may be an interesting case in point. Child behavior may also help channel adult instruction in novel directions. In research on the "why" questions that preschoolers pose to parents, Callanan and Oakes (1992) found that children frequently asked about how the world works ("How does the telephone know which house to call?") or why something is the way it is ("Why is the sky blue?"). These questions tended to initiate shared activity that included explanations by adults that were responsive to the interests of the child. And as the question about the telephone suggests, a child's interests may direct adult attention to aspects of the environment that are often overlooked or taken for granted by the more experienced partner.

Resistance may also be a form of adult–child transaction that contributes to cultural change. Although there is research on resistance among adults as a source of change (Scott, 1985), studies on resistance by children as a cultural force are lacking. (However, see chap. 13, this volume, for research on children's resistance from the perspective of individual development.) Research on resistance in adults is largely concerned with the distribution of power within a culture. This issue may not seem relevant to developmental study, though Ogbu's (1994) research on the devaluing of academic achievement among African American youth is interesting to consider in this light.

To summarize, social processes critical to cognitive development offer opportunity for cultural values and practices to be transmitted and changed. Most important aspects of human development are overdetermined. In like fashion, important elements of culture, such as ways of supporting cultural maintenance and cultural change across generations, may also be overdetermined in the social processes that support and guide cognitive development.

CONCLUSIONS

Social transactions with people and with the signs and tools of culture, in conjunction with biological capabilities, form the bases of cognitive development. These transactions are essential to the organization and adaptation of the individual as well as integral to the sustainability of culture. Examination of the transactions between cognitive development and culture across generations may be useful to theories of cognitive development on several fronts. First, these transactions demonstrate the active role that children play in maintaining and changing culture. Second, they suggest ways that cultural change can emerge from normative developmental processes. Third, they reveal how characteristics of childhood provide unique contributions to cultural change. And, fourth, they reveal how social processes of cognitive development contribute to culture across generations.

It is important to stress that the cross-generational data required to support these claims are difficult to come by. However, relevant information may be available in contemporary developmental studies as well as historical analysis. For instance, it may be fruitful to examine social processes of cognitive development from the vantage point of how these processes contribute to maintaining or changing conventional cultural practices. Some ideas along these lines are suggested in this chapter. It also may be useful to examine instances of major change, both in terms of human development and culture. Some periods of cognitive developmental change that may be particularly relevant are toddlerhood, the time when a primary social–cultural–cognitive form, language, enters the scene; the 5-to-7 shift, known as the age of reason and responsibility, may also be the first point in childhood when the adoption or rejection of conventional cultural forms takes on a serious cast (Sameroff & Haith, 1996); and early adolescence may be the time when relations with the family, peers, and community undergo substantial revision (Schlegel & Barry, 1991). As White (1996) pointed out, these developmental transitions do not represent maturation per se; he suggested looking at a transition as a statement about the relationship between the growing child and his or her society (p. 27).

Historical situations at times of major cultural change may also provide evidence. Of particular interest are times and places when major shifts in social and technological support for thinking occurred, for example, the establishment of formal schooling in a culture (Tyack, 1974), the adoption of Arabic numerals during the Renaissance (Swetz, 1987), changes in literacy practices associated with computer technology (Wallace, 1999), and shifts in practices of measurement (Crosby, 1997). Each of these cultural changes led to many changes in human experience, including social interactions and, presumably, cognitive development. A significant stumbling block in this type of analysis, especially with more remote historical instances, is the lack of detail about the experiences of children during periods of change. Major developments in computer technology, especially in the context of globalization, may be important to this type of analysis.

Although it was useful for the present discussion to treat cultural maintenance and change as separate forces, in actuality the distinction between these aspects of culture is much less sharp. Moreover, human behavior and innovations that foster change may be constrained by conventional ways of interacting and of using cultural tools. In this light, a broader definition of innovation or creativity may be useful. Many definitions of creativity imply that a creative act is completely novel, that is, there is no preexisting form. This conception of creativity fails to recognize that the use of conventional forms toward new ends is innovative. In fact, this may be one of the primary ways that innovation contributes to cultural change (Tomasello et al., 1993). Of course, it is worth remembering that human beings are not particularly good at this type of innovation. The organized, goal-directed nature of human activity and its intricate relation to cultural tools and practices, previously discussed, can lead to difficulty in identifying and solving new problems, as

Duncker (1945) observed in his research on functional fixedness. It may be that different cognitive processes underlie behaviors that coincide with cultural conventions versus those that depart from these conventions. Although this idea represents a different level of analysis than the focus of this chapter, it has important implications for the issues discussed here.

Finally, it is important to remember that two types of change have been discussed in this chapter. Cognitive development, a process of individual change, has been examined in relation to cultural change, a group process. Developmental research is largely concerned with the former, and when developmentalists examine cultural change, they tend to concentrate on how cultural change impacts children in a particular setting at a particular point in time. Here I argue that cognitive development and cultural change are intertwined and that their relation is active, prospective, and transactional. In other words, the ways in which a culture maintains its identity and creates change across generations are intricately tied to the social context of cognitive development. Better understanding of this complex, dynamic, and open-ended process will be aided by the view of development represented in the sociocultural perspective and the transactional model. Research along these lines will advance understanding of the relations among cognitive development, social processes, and culture, which may, in turn, illuminate how children contribute to and help sustain culture.

REFERENCES

Altman, I., & Rogoff, B. (1987). World views in psychology: Trait, interactional, organismic, and transactional perspectives. In D. Stokols & I. Altman (Eds.), *Handbook of environmental psychology* (Vol. 1, pp. 7–40). New York: Wiley.

Bakeman, R., & Adamson, L. B. (1984). Coordinating attention to people and objects in mother–infant and peer–infant interaction. *Child Development, 55,* 1278–1289.

Bandura, A. (1986). *Social foundations of thought and action: A social cognitive theory.* Englewood Cliffs, NJ: Prentice-Hall.

Barkow, J. H., Cosmides, L., & Tooby, J. (1992). *The adapted mind: Evolutionary psychology and the generation of culture.* New York: Oxford University Press.

Barnett, H. G. (1953). *Innovation: The basis of cultural change.* New York: McGraw-Hill.

Berry, J. W. (2007). Acculturation. In J. E. Grusec & P. D. Hastings (Eds.), *Handbook of socialization: Theory and research* (pp. 543–558). New York: Guilford Press.

Bjorklund, D. F. (2007). *Why youth is not wasted on the young: Immaturity in human development.* Malden, MA: Blackwell.

Bjorklund, D. F., & Pelligrini, A. D. (2002). *The origins of human nature: Evolutionary developmental psychology.* Washington, DC: American Psychological Association.

Bronfenbrenner, U., & Morris, P. (2006). The ecology of developmental processes. In W. Damon & R. W. Lerner (Series Eds.) & R. M. Lerner (Vol. Ed.), *Handbook of child psychology: Vol 1. Theoretical models* (pp. 793–828). New York: Wiley.

Callanan, M. A., & Oakes, L. M. (1992). Preschoolers' questions and parents' explanations: Causal thinking in everyday activity. *Cognitive Development, 7*, 213–233.

Campos, J. J., & Stenberg, C. R. (1981). Perception, appraisal, and emotion: The onset of social referencing. In M. E. Lamb & L. R. Sherrod (Eds.), *Infant social cognition: Empirical and theoretical considerations* (pp. 273–314). Hillsdale, NJ: Erlbaum.

Carlton, J. (1997). *Apple: The insider story of intrigue, egomania, and business blunders*. New York: Random House.

Cassell, J. (2004). Towards a model of technology and literacy development: Story listening systems. *Applied Developmental Psychology, 25*, 75–105.

Chavajay, P., & Rogoff, B. (1999). Cultural variation in management of attention by children and their caregivers. *Child Development, 35*, 1079–1090.

Cole, M. (1996). *Cultural psychology: A once and future discipline*. Cambridge, MA: Harvard University Press.

Crosby, A. W. (1997). *The measure of reality: Quantification and western society, 1250–1600*. Cambridge, England: Cambridge University Press.

D'Andrade, R. G. (1981). The cultural part of cognition. *Cognitive Science, 5*, 179–195.

Duncker, K. (1945). On problem solving. *Psychological Monographs, 58*(5, Whole No. 270).

Elman, J. L., Bates, E. A., Johnson, M. H., Karmiloff-Smith, A., Parisi, D., & Plunkett, K. (1996). *Rethinking innateness: A connectionist perspective on development*. Cambridge, MA: MIT Press.

Erikson, E. H. (1963). *Childhood and society*. New York: Norton.

Fagot, B. I. (1977). Consequences of moderate cross-gender behavior in preschool children. *Child Development, 48*, 902–907.

Fiese, B. H. (2006). *Family routines and rituals*. New Haven, CT: Yale University Press.

Gauvain, M. (1995). Thinking in niches: Sociocultural influences on cognitive development. *Human Development, 38*, 25–45.

Gauvain, M. (2001). *The social context of cognitive development*. New York: Guilford Press.

Gauvain, M., Altman, I., & Fahim, H. (1983). Homes and social change: A cross-cultural analysis. In N. R. Feimer & S. Geller (Eds.), *Environmental psychology: Directions and perspectives* (pp. 180–218). New York: Praeger.

Gauvain, M., de la Ossa, J., & Hurtado, M. (2001). Parental guidance as children learn to use cultural tools: The case of pictorial plans. *Cognitive Development, 16*, 551–575.

Gauvain, M., & Perez, S. M. (2007). The socialization of cognition. In J. E. Grusec & P. D. Hastings (Eds.), *Handbook of socialization: Theory and research* (pp. 588–613). New York: Guilford Press.

Goodnow, J. J. (1990). The socialization of cognition: What's involved? In J. W. Stigler, R. A. Shweder, & G. Herdt (Eds.), *Cultural psychology: Essays on comparative human development* (pp. 259–286). Cambridge, England: Cambridge University Press.

Goodnow, J. J., Miller, P. J., & Kessel, F. (1995). *Cultural practices as contexts for development*. San Francisco: Jossey-Bass.

Goody, J. (1977). *The domestication of the savage mind*. Cambridge, England: Cambridge University Press.

Gottlieb, G. (1992). *Individual development and evolution: The genesis of novel behavior*. New York: Oxford University Press.

Greenfield, P. M. (2004). *Weaving generations together: Evolving creativity in the Maya of Chiapas*. Santa Fe, NM: SAR Press.

Greenfield, P. M., & Lave, J. (1982). Cognitive aspects of informal education. In D. A. Wagner & H. W. Stevenson (Eds.), *Cultural perspectives on child development* (pp. 181–207). San Francisco: Freeman.

Grusec, J. E., & Hastings, P. D. (2007). *Handbook of socialization: Theory and research*. New York: Guilford Press.

Hinde, R. A. (1987). *Individuals, relationships, and culture*. Cambridge, England: Cambridge University Press.

Inkeles, A., & Smith, D. H. (1974). *Becoming modern: Individual change in six developing countries*. Cambridge, MA: Harvard University Press.

Kagitçibasi, C. (2007). *Family and human development across cultures: A view from the other side* (2nd ed.). Mahwah, NJ: Erlbaum.

Kuczynski, L. (2003). *Handbook of dynamics in parent–child relations*. Thousand Oaks, CA: Sage.

Lancy, D. F. (1996). *Playing on the mother-ground: Cultural routines for children's development*. New York: Guilford Press.

Lave, J. (1988). *Cognition in practice*. Cambridge, England: Cambridge University Press.

Lave, J., & Wenger, E. (1991). *Situated learning: Legitimate peripheral participation*. Cambridge, England: Cambridge University Press.

LeVine, R., Dixon, S., LeVine, S., Richman, A., Leiderman, P. H., Keefer, C. H., et al. (1994). *Child care and culture: Lessons from Africa*. Cambridge, England: Cambridge University Press.

Lightfoot, C. (1997). *The culture of adolescent risk-taking*. New York: Guilford Press.

Maccoby, E. E. (1992). The role of parents in the socialization of children: An historical overview. *Developmental Psychology, 28*, 1006–1017.

Merriam-Webster's Collegiate Dictionary (11th ed.). (2003). Springfield, MA: Merriam-Webster, Inc.

Nelson, K. (2007). *Young minds in social worlds*. Cambridge, MA: Harvard University Press.

Ogbu, J. U. (1994). From cultural differences to differences in cultural frame of reference. In P. M. Greenfield & R. R. Cocking (Eds.), *Cross-cultural roots of minority child development* (pp. 365–391). Hillsdale, N J: Erlbaum.

Olson, D. R. (1994). *The world on paper: The conceptual and cognitive implications of writing and reading*. Cambridge, England: Cambridge University Press.

Rogoff, B. (1990). *Apprenticeship in thinking: Cognitive development in social context*. New York: Oxford University Press.

Rogoff, B. (2003). *The cultural nature of human development*. Oxford, England: Oxford University Press.

Rommetveit, R. (1985). Language acquisition as increasing linguistic structuring of experience and symbolic behavior control. In J. V. Wertsch (Ed.), *Culture, communication, and cognition* (pp. 183–204). Cambridge, England: Cambridge University Press.

Sameroff, A. J. (1989). General systems and the regulation of development. In M. R. Gunnar & E. Thelen (Eds.), *Systems and development* (Vol. 22, pp. 219–235). Hillsdale, NJ: Erlbaum.

Sameroff, A. J., & Haith, M. M. (Eds.) (1996). *The five to seven year shift: The age of reason and responsibility*. Chicago: The University of Chicago Press.

Schlegel, A., & Barry H, III, (1991). *Adolescence: An anthropological enquiry*. New York: Free Press.

Schiefflin, B. B., & Ochs, E. (1986). *Language socialization across cultures*. Cambridge, England: Cambridge University Press.

Scott, J. C. (1985). *Weapons of the weak*. New Haven, CT: Yale University Press.

Serpell, R. (1993). Interface between sociocultural and psychological aspects of cognition: A commentary. In E. Forman, N. Minick, & A. Stone (Eds.), *Contexts for learning* (pp. 357–368). Oxford, England: Oxford University Press.

Serpell, R. (1995). Situated theory as a bridge between psychology, history, and educational practice. In L. M. W. Martin, K. Nelson, & E. Tobach (Eds.), *Sociocultural psychology* (pp. 21–42). Cambridge, England: Cambridge University Press.

Serpell, R., & Hatano, G. (1997). Education, schooling, and literacy. In J. W. Berry, Y. H. Poortinga, & J. Pandey (Eds.), *Handbook of cross-cultural psychology: Vol. 3. Basic processes and human development* (2nd ed., pp. 339–376). Needham Heights, MA: Allyn & Bacon.

Subrahmanyam, K., Smahel, D., & Greenfield, P. (2006). Connecting developmental constructions to the Internet: Identity presentation and sexual exploration in online teen chat rooms. *Developmental Psychology, 42*, 395–406.

Steinberg, L. (2007). Risk taking in adolescence: New perspectives from brain and behavioral science. *Current Directions in Psychological Science, 16*, 55–59.

Swetz, F. J. (1987). *Capitalism and arithmetic*. LaSalle, IL: Open Court.

Tomasello, M., Carpenter, M., Call, J., Behne, T., & Moll, H. (2005). Understanding and sharing of intentions: The origins of cultural cognition. *Behavioral and Brain Sciences, 28*, 675–735.

Tomasello, M., Kruger, A. C., & Ratner, H. H. (1993). Cultural learning. *Behavioral and Brain Sciences, 16*, 495–511.

Tudge, J. R. H. (1992). Processes and consequences of peer collaboration: A Vygotskian analysis. *Child Development, 63*, 1364–1379.

Turkle, S. (1984). *The second self: Computers and the human spirit*. New York: Touchstone.

Tyack, D. B. (1974). *The one best system: A history of American urban education*. Cambridge, MA: Harvard University Press.

Valsiner, J., & Lawrence, J. (1997). Human development in culture across the life span. In J. W. Berry, Y. H. Poortinga, & J. Pandey (Eds.), *Handbook of cross-cultural psychology: Vol. 3. Basic processes and human development* (2nd ed., pp. 69–106). Needham Heights, MA: Allyn & Bacon.

Vygotsky. L. S. (1981). The genesis of higher mental functions. In J. V. Wertsch (Ed.), *The concept of activity in Soviet psychology* (pp. 144–188). Armonk, NY: Sharpe.

Vygotsky, L. S. (1987). *The collected works of L. S. Vygotsky: Vol. 1. Problems of general psychology*. New York: Plenum Press.

Wallace, P. (1999). *The psychology of the internet*. Cambridge, England: Cambridge University Press.

Wertsch, J. V. (1998). *Mind as action*. New York: Oxford University Press.

White, S. H. (1996). The child's entry into the "age of reason." In A. J. Sameroff & M. M. Haith (Eds.), *The five to seven year shift: The age of reason and responsibility* (pp. 17–30). Chicago: The University of Chicago Press.

Wood, D. J., & Middleton, D. (1975). A study of assisted problem solving. *British Journal of Psychology, 66*, 181–191.

Zigler, E. F., & Hall, N. W. (2000). *Child development and social policy: Theory and applications*. New York: McGraw-Hill.

10

THE TRANSITION TO SCHOOL: CHILD-INSTRUCTION TRANSACTIONS IN LEARNING TO READ

FREDERICK J. MORRISON AND CAROL McDONALD CONNOR

"The development of the child is a product of the continuous dynamic interactions of the child and the experience provided by his or her family and social context *(including school)* [italics added]" (Sameroff & MacKenzie, 2003, p. 614). As this quote illustrates, the transactional view, although recognizing the role of the larger sociocultural milieu in shaping developmental trajectories, has emphasized the family context in theory and research on the complex interplay of child and environment in human development. This focus naturally highlights the varying patterns of continuous interaction that characterize most parent–child relations across significant time periods in the child's life. What about transactional discontinuities in development over those same periods in which the child is regularly exposed to time-varying environments on a systematic basis? This is the common question that defines schooling in the United States and most of the world. Starting in kindergarten (and

We thank the members of the Pathways to Literacy; the Individualizing Student Instruction Projects; and the children, parents, teachers, and administrators who participated in the research presented. Preparation of this chapter was supported by grants from the National Institute for Child Health and Human Development (HD27176 to F. J. Morrison and R01HD48539 to C. M. Connor) and the U.S. Department of Education, Institute for Education Sciences (R305H04013 and R305B070074 to C. M. Connor).

increasingly earlier) children interact with adults (teachers, aides) for large numbers of hours each week for about 10 months a year. In most cases, after a summer break they subsequently interact for the next 10 months with a completely different person who, in many instances, has very little knowledge of their skills or other aspects of their personality. What impact does this discontinuity have on the developing academic skills of children?

In this chapter we describe our ongoing research on the impact of instruction on children's literacy development in the early elementary school years. Our findings have revealed that the most effective instruction for children learning to read varies widely and depends in large part on the initial skill level of the child. Furthermore, the development of reading skills across grades varies significantly depending on the degree of match between the child's skill level and the subsequent pattern of instruction. In other words, the transactional discontinuity across grades holds major consequences for the child's literacy development.

BEYOND THE READING WARS

There has been a long-standing controversy regarding the best way to teach children how to read (Ravitch, 2001). The debate has centered on the efficacy of phonics or code-based instruction versus whole-language or meaning-based instruction (Rayner, Foorman, Perfetti, Pesetsky, & Seidenberg, 2001). Code-based instruction concentrates on helping children learn to crack the code of reading—letters have sounds that combine to make words in fairly predictable ways. Meaning-based instruction views learning to read as a more natural process (Goodman, 1970) that requires abundant experiences with books and other print within a literature-rich environment (Dahl & Freppon, 1995). However, in contrast to the polarized views, mounting evidence shows that most children develop stronger reading skills when they are provided with explicit decoding instruction in combination with meaningful reading activities—a more balanced approach (Cunningham & Hall, 1998; Guthrie, Schafer, & Huang, 2001; Pressley, 1998; Rayner et al., 2001; Taylor, Pearson, Clark, & Walpole, 2000). But even a balanced approach to teaching reading leaves in place an implicit assumption that a curriculum or even a specific instructional practice will be equally effective for all children (National Reading Panel [NRP], 2000; Ross, Smith, Slavin, & Madden, 1997). Accumulating research indicates that the efficacy of a particular instructional practice will depend on the skill level of the student. Instructional strategies that help one student (e.g., with weaker vocabulary skills) may be ineffective for another student (e.g., with stronger vocabulary skills). This phenomenon has been called Aptitude × Treatment interaction (Cronbach & Snow, 1977) or, more recently, Child × Instruction interactions (Connor, Jakobsons, Crowe, & Granger, in press; Connor, Morrison, & Katch, 2004; Connor, Morrison, & Petrella, 2004; Juel & Minden-Cupp, 2000).

DIMENSIONS OF INSTRUCTION

One challenge in observational and experimental research investigating individualized instruction is conceptualizing the complexities of students' classroom experiences in ways that are meaningful for teachers while also permitting statistical analysis of their relations to student outcomes. Much of reading research has compared the relative effectiveness of two broad curriculum types, meaning-focused and skill- or code-based instruction (Rayner et al., 2001). However, during any given school day, teachers may provide a variety of activities that incorporate both of these major curriculum types. The teacher may use code- or skill-based instruction for teaching phonological skills and a meaning-focused approach as he or she sets up the library corner and encourages students to read independently. Comparing one curriculum with another may overlook the variety and complexity of learning activities students actually experience, including who (teacher or child) manages the activity and how instruction shifts over the course of the year. Additionally, accumulating evidence indicates that literacy is multidimensional (Hoover & Gough, 1990; Rayner et al., 2001; Senechal & LeFevre, 2001) rather than a more global construct. Literacy is distinct from (though connected to) oral language and metalinguistic awareness (Mason & Stewart, 1990; Whitehurst & Lonigan, 1998). Different components of the home literacy environment (formal vs. informal emergent literacy experiences) differentially predict the distinct components of literacy (Senechal & LeFevre, 2002; Senechal, LeFevre, Thomas, & Daley, 1998; Storch & Whitehurst, 2002). For example, reading to children promotes primarily vocabulary and oral language skills, whereas direct instruction in letter sound knowledge and word decoding promotes those skills without much effect on vocabulary. As a parallel argument, if literacy has multiple dimensions, then examining sources of classroom influence on literacy multidimensionally at the level of the individual child should prove more informative than examining the impact of these dimensions globally at the curriculum or classroom level. In our work, we have organized the content of classroom literacy activities across five dimensions: (a) code focused versus meaning focused, (b) teacher versus teacher/child versus peer versus child managed, (c) explicit versus implicit, (d) student level versus classroom level, (e) and change in instruction over the school year.

Code- Versus Meaning-Focused Activities

The code- versus meaning-focused dimension captures the content focus of language and literacy activities. Code-focused activities (Foorman, Francis, Fletcher, Schatschneider, & Mehta, 1998; Rayner et al., 2001) include teaching children how to name and write letters, rhyme words (Torgesen, Burgess, Wagner, & Rashotte, 1994; Torgesen et al. 1999), relate letters to the sounds they make, and sound out words (*phonological decoding*). In contrast, activities designed to help students understand words and passages, comprehend what is read to them and what they read, and enhance receptive and expressive language skills including listening comprehension (Scarborough, 1990) are considered meaning focused (Dahl & Freppon, 1995; Foorman et al., 1998; Juel & Minden-Cupp, 2000). See Table 10.1.

Teacher-Managed Versus Teacher/Child-Managed Versus Peer-Managed Versus Child-Managed Activities

One important element of classroom learning reflects who is focusing the child's attention—the teacher, the peers, or the student (Connor, Morrison, & Katch, 2004; Connor et al., 2004b; Morrison, Bachman, & Connor, 2005)—or whether the attention is jointly focused (teacher and students interacting). These studies, conducted with preschool and first-grade children, suggest that those with higher levels of vocabulary benefit most from more child-managed learning, whereas those with developing skills profit from teacher- and teacher/child-managed experiences. In the framework used in this study, child-centered or child-initiated learning may be teacher managed, child managed, or managed jointly in a teacher/child-managed situation; what is at issue is who is directing the child's attention—the teacher and/or the child? For example, a teacher reading a book to students without discussion, even if the children selected the book, would be considered teacher-managed because the teacher is focusing the children's attention and doing most of the talking. Sharing, scaffolding, or interactive read alouds are considered to be teacher/child managed within this framework because the teacher is actively

TABLE 10.1
Observed Instructional Activities Organized by Dimensions of Instruction

Instruction dimension	Teacher/child managed	Child/peer managed
Code focused	■ Alphabet activities ■ Letter sight and sound ■ Phonological awareness, onset rhyme, blending and segmenting ■ Word segmentation	■ Spelling ■ Phonics worksheets, handwriting activities ■ Sight word flash cards ■ Computers—phonological awareness, and so on
Meaning focused	■ Vocabulary ■ Teacher read aloud ■ Student read aloud ■ Writing instruction ■ Listening comprehension ■ Discussion ■ Comprehension strategies	■ Buddy reading ■ Sustained silent reading ■ Reading comprehension worksheets ■ Student individual writing ■ Books on CD ■ Computers—meaning ■ Independent writing

involved with and responsive to the child. In contrast, activities in which the child is working with peers without immediate teacher guidance, such as playing a phonics game or reading with a peer, are considered peer managed. In one recent study (Connor, Morrison, & Slominski, 2006) we found that organized peer play activity significantly enhanced the vocabulary growth of preschool children who started the school year with relatively low vocabulary skills. Finally, independent individual activities, such as working alone to complete a journal or workbook page, are considered child managed given that the student is directing his or her own attention without the support of others.

Explicit Versus Implicit Dimension

The explicit versus implicit dimension (Connor, Morrison, & Petrella, 2004; Connor, Morrison, & Slominski, 2006) incorporates the idea that activities can be centrally or incidentally focused on promoting a specific outcome (in this study, specific components of literacy). This dimension is defined relative to the outcome being explored. For example, if the target outcome is reading comprehension, activities that directly focus on the extraction and construction of meaning from text (Snow, 2002), such as teaching comprehension strategies (NRP, 2000), would be considered explicit. During a book reading, a teacher might read text aloud to children, which would represent an implicit comprehension activity. This kind of activity might be expected to build comprehension skills, but it would do so implicitly rather than systematically and explicitly (Connor et al., 2004b). Weaving in instruction about testing one's understanding while reading or looking up unfamiliar words in the dictionary would represent explicit comprehension instruction.

Classroom- Versus Student-Level Dimension

The classroom- versus student-level dimension considers the extent to which instruction is the same or different for each child in the classroom. Literacy activities may be at the classroom level. To be considered classroom-level instruction, the students are all doing substantially the same thing at the same time (e.g., completing the same phonics activity). For instance, the teacher could be reading aloud to the entire class or the children could be working individually or in small groups (e.g., centers). In contrast, with student-level instruction activities, children are engaged in substantially different activities at the same time. Teachers may provide student-level instruction in small groups or they may work with children individually (e.g., centers with different activities, tutoring one child while the rest do other activities). A video-coding system developed in our labs permits coding of each individual student's participation in specific literacy activities so that we can examine this dimension of instruction.

Note that specific types of activities (e.g., book reading) can occur on both the classroom and student levels. For example, in one classroom a teacher might read aloud to the entire class (teacher managed, meaning focused, implicit, classroom level), whereas in another, a teacher might read aloud to a small group of children (teacher managed, meaning focused, implicit, child level) while the rest of the children engage in substantially different activities, such as writing in their journals (child managed, meaning focused, explicit, child level).

Change in Amount of Instruction Over the School Year

The final dimension cuts across the previous four and highlights changes in instructional activities over the school year. One provocative finding in the Juel and Minden-Cupp (2000) study revealed that some teachers changed their instructional emphasis over the course of the first-grade school year. For example, one teacher began the year with a strong focus on explicit, teacher-managed decoding instruction that tapered off as the year progressed and as children mastered basic skills. In this class, children with weaker fall reading skills (i.e., children in the low reading group) achieved stronger spring decoding scores than did children in the low reading group in other classrooms with less initial teacher-managed decoding focus. In our work (Connor, Morrison, & Katch, 2004), we also found that for first graders with low initial vocabulary scores, small amounts of child-managed, meaning-focused (e.g., independent reading) activities at the beginning of the school year followed by steady increases as the year progressed was optimal. The opposite pattern held for children with initially high scores for whom steady amounts of child-managed, meaning-focused instruction were most beneficial.

Putting the Dimensions Together

A key feature of these dimensions is that they operate simultaneously. Thus, activities might be fundamentally designated as teacher, teacher/child, peer, or child managed but then assigned three other modifiers to reflect the focus of the content (code vs. meaning), the nature of the content (explicit vs. implicit), and the level (classroom vs. student) at which it operates. Together, these dimensions can be combined to produce a total of 16 possible designations.

CHILD FACTORS

Notwithstanding the importance of the instructional environment and other teaching dimensions, children's characteristics play independent and interactive roles in shaping literacy trajectories during the early elementary years. Accumulating evidence reveals that a number of literacy skills in the early grades consistently predict children's later reading and academic success, including alphabet knowledge, phonological awareness, letter and word recognition, and phonological decoding (National Institute of Child Health and Development Early Child Care Research Network [NICHD ECCRN], 2005; Poe, Burchinal, & Roberts, 2004; Rayner et al., 2001; Scarborough, 1998; Schatschneider, 2004; Snow, Burns, & Griffin, 1998). Language skills, particularly vocabulary and metalinguistic awareness, are also consistent predictors of later reading success, especially as comprehending what is read becomes important (Anderson & Freebody, 1981; Loban, 1976; NICHD ECCRN, 2005; NRP, 2000; Poe et al., 2004; Scarborough, 1990, 2001; Snow et al., 1998; Storch & Whitehurst, 2002). In contrast to early decoding skills, growth in vocabulary may be highly stable and resistant to intervention and the effect of schooling (Hart & Risley, 1995; Morrison, Smith, & Dow-Ehrensberger, 1995). However, rich meaning-focused literacy experiences, such as storybook reading both at home (Beals & DeTemple, 1993; Senechal et al., 1998) and in school (Beck, McKeown, & Kucan, 2005; Dickinson, Anastasopolous, McCabe, Peisner-Feinberg, & Poe, 2003; Dickinson & Tabors, 2001), can contribute to children's language growth.

In addition, self-regulation plays a central role in early literacy development and academic success (Morrison et al, 2005). Students with weak self-regulation present classroom management challenges and demonstrate less growth in reading skills than do children with stronger self-regulation (Cameron, Connor, Morrison, & Jewkes, 2008; McClelland et al., 2007). Typically, students' social and behavioral regulation has been assessed using teacher or parent report, with clear limitations. Recently, research in executive functioning and the specific traits associated with school performance—working memory, attention, and task inhibition and switch—have offered ways to test behavioral self-regulation directly (Cameron, McClelland, et al., 2008). Students are first taught to follow simple commands: "Touch your head," and "touch your toes." They are then asked to do the opposite, "When I say touch your head, I want you to touch your toes." In a recent intervention study we investigated the effect of first graders' behavioral self-regulation on their literacy development, along with the effect of teachers' planning, classroom management, and instruction on students' self-regulation and literacy skill growth. Preschoolers' performance on the head–toes task predicts emergent literacy, mathematics, and vocabulary growth (McClelland et al., 2007).

Using a more complicated form of the head–toes task, such as adding knees and shoulders, we found that first graders who had had stronger self-regulation in the fall generally demonstrated significantly stronger reading and vocabulary skill growth than did children with weaker fall skills. The effect of students' self-regulation on their language and literacy was over and above the effect of their initial skills and teachers' classroom practice. Plus, teachers' actions in the classroom affected students' self-regulation growth and literacy development. Specifically, teachers' planning and instruction were positively related to students' self-regulation and literacy skill growth. Poor classroom management, the increasing time students spent in transitions between activities and other noninstructional activities, was negatively related to their literacy skill growth, but only when students' self-regulations skills were considered as a covariate. One might conjecture that teachers' planning and classroom management can help children compensate for weaker self-regulation, resulting in stronger literacy learning or perhaps helping to strengthen self-regulation.

CHILD × INSTRUCTION INTERACTIONS

In recent work, we have examined Child × Instruction interactions in first grade, second grade (Connor, Morrison, & Underwood, 2007), third grade (Connor, Morrison, & Petrella, 2004), and most recently preschool (Connor, Morrison, & Slominski, 2006). Other work by Connor and colleagues using a Reading First sample and the same dimensions-of-instruction framework examined Child × Instruction interaction effects on reading comprehension over the first 3 years of elementary school (Connor et al., in press). We have focused on this age range (3 years to third grade, or the school transition period) for a variety of reasons. First, it has become clear that meaningful individual differences in children's cognitive, language, literacy, and social skills emerge well before formal schooling begins in kindergarten or first grade (Morrison et al., 2005). Second, without substantial intervention, these differences persist well into elementary school. Hence, it has become increasingly apparent to us that fuller understanding of children's academic trajectories will benefit from simultaneous consideration of the instruction they receive in conjunction with the level of academic skills they bring to the classroom.

Two examples illustrate the pattern of findings emerging from this work. In a first-grade study (Connor, Morrison, & Katch, 2004), we followed children across the academic year measuring a range of literacy and language skills in the fall and spring and systematically observing their classrooms for a whole day three times during the year. We found that for children who started the year with relatively low word-decoding skills, more time spent in teacher-managed, code-focused instruction yielded greater reading gains by spring. In contrast, for students with higher initial decoding skills, spending more time in teacher-managed, code-focused instruction made no difference. For students with stronger vocabularies, though, more time in child-managed, meaning-focused activities yielded greater improvement over the year, although more time produced less improvement for children with initially lower vocabulary scores. We followed these children into third grade, focusing on growth of reading comprehension skills (Connor et al., 2004b). Here we found that for children starting with lower comprehension scores, more time in teacher-managed explicit instruction in comprehension strategies (e.g., summarizing and inferring) was associated with greater gains. For students with higher initial skills, more teacher-managed explicit instruction had no effect, whereas greater amounts of child-managed explicit activities significantly improved growth for these children.

Moreover, there is evidence that these Child × Instruction interactions are causally implicated in the widely varying achievement observed within and across classrooms and schools. In a randomized control field trial, instruction was manipulated such that randomly assigned teachers were taught how to individualize instruction, taking into account Child × Instruction interactions. The comparison group provided high-quality evidence-based instruction but did not consider Child × Instruction interactions. In addition to a significant effect of treatment (Connor, Morrison, Fishman, Schatschneider, & Underwood, 2007), when students were observed in the classroom, we found that total amounts of instruction received did not predict their literacy growth. Nevertheless, the precision with which teachers taught amounts and types of instruction on the basis of Child × Instruction interactions did positively affect students' literacy skill growth. Overall, across a broad spectrum of grades (from preschool to third grade) and skills (decoding, letter and word reading, vocabulary, fluency, and comprehension), Child × Instruction interactions emerged as a pervasive feature of early literacy development.

FROM INTERACTION TO TRANSACTION: INSTRUCTIONAL EFFECTS ACROSS GRADES

The results from these naturalistic observation studies, although compelling, are descriptive and describe mean performance among children. Thus, the question is, do we see the similar effects when we examine scores for real children in actual classrooms? From the data set for the first-grade study we selected two children, Aaron and Christopher (names have been changed), whose vocabulary and word-reading skills were below standardized norms and their classroom means. Aaron was in a classroom in which the teacher was observed to provide about 10 minutes per day of teacher-managed, code-focused instruction and about 12 minutes per day of child-managed, meaning-focused instruction in the fall, increasing to almost 30 minutes by spring. In contrast, Christopher was in a classroom in which the teacher provided less than 4 minutes of teacher-managed, code-focused instruction and a steady dose of child-managed meaning-focused instruction (between 30 and 40 minutes) all year long. According to our models, Aaron received a more optimal pattern of instruction than did Christopher. Examining their individual growth, we discovered that Aaron demonstrated about a years' worth of growth in word-reading skills from fall to spring, whereas Christopher demonstrated less than half that amount of growth in basic reading skills.

The transactional nature of instruction is further illustrated as we follow Aaron into second grade. Although Aaron made substantial gains in first grade, these gains were not enough to bring his skills in line with his peers—he was still reading well below grade level at the beginning of second grade. Following a transactional model, students' reading skills should influence the impact of the amount and type of instruction they receive in second grade—again there should be Child × Instruction interactions.

Less is known about the nature and impact of language arts instruction in second grade. However, second grade may be as important for young readers as is first grade. Reading skill growth trajectories for students entering first grade with reading skills falling below the 30th percentile (similar to Christopher and Aaron) were examined (Spira, Bracken, & Fischel, 2005). Fully one third of these children demonstrated significant growth in reading by the end of second grade and continued to improve through fourth grade, on average achieving grade level scores. However, the two thirds who ended second grade with reading skills below the 30th percentile failed to achieve grade level reading by the end of third grade, and on average, their scores fell below the 10th percentile by the end of fourth grade.

A review of the literature on second-grade classroom instruction indicates that less time is spent on explicit code-focused instruction in second grade when compared with first. Foorman et al. (2006), for example, observed language arts blocks in first- and second-grade classrooms. More than 25% of this time included explicit, code-focused instruction in first grade but less than 15% in second grade. In the same way, site visits in a group of Florida Reading First schools revealed that the amount of code-focused instruction decreased 50% in second grade compared with first (Connor et al., in press). The same trend was seen following our sample from first grade to second grade (Connor, Morrison, & Underwood, 2007). Substantially less time was spent in teacher-managed, code-focused instruction in second grade compared with first grade.

The assumption on the part of educators and researchers appears to be that code-focused instruction is more important in first grade than in second grade. Generally, research supports these assumptions. For example, results from the NRP meta-analysis (McCardle & Chhabra, 2004; NRP, 2000) revealed that the overall effect size for phonics instruction was moderate to large for children in kindergarten and first grade (.55). For children in second through sixth grade, mean effect size was smaller (.27). When prior achievement was considered, phonics instruction had no significant effect size for children with low reading skills in second through sixth grade but a large effect size (.74) for first-grade children considered at risk for academic underachievement.

Foorman et al. (2006) reported that students with teachers who were rated as more competent on a scale that examined planning, management, instruction, monitoring student learning, and knowledge about reading instruction and who spent more time engaged in phonemic awareness and alphabet instruction than in nonreading activities achieved stronger word-attack skills in first grade. In contrast, for second-grade students, the opposite—less phonemic awareness and more nonreading time—was associated with stronger word-attack scores. This, the authors suggested, was because "presumably, phonemic awareness and alphabet activities are no longer critical elements of instruction in second grade and, in this sample, accounted for only 5% of instructional time" (p. 22).

Transactional theory would suggest otherwise. For some children, the impact of instruction may depend on the instruction children received in first grade inasmuch as this instruction affected students' reading skill growth and thus the skills they brought to the second-grade classroom. In turn, the effect of instruction received in second grade would depend on students' language and literacy skills. Second-grade instruction would build on the instruction received in first grade. Indeed, Sanders and Horn (1998) found that the impact of effective math teaching was cumulative for students in third through eighth grades. Students assigned to

> ineffective [math teachers] continue[ed] to show the effects of such teachers even when these students [were] assigned to very effective teachers in subsequent years. . . . the teacher effects are both additive and cumulative with little evidence of compensatory effect of more effective teachers in later grades. (Sanders & Rivers, 1996, cited in Sanders & Horn, 1998, p. 254)

However, following the children in our first-grade study (Connor, Morrison, & Katch, 2004) into second grade was challenging because longitudinal examination of instruction effects on children's reading growth requires that data have a cross-classified structure. Children attend one set of classrooms in first grade and different classrooms with different teachers and peers in second grade. Fortunately, analytic strategies have been developed to accommodate cross-classified data structures—cross-classified random effects models (Raudenbush & Bryk, 2002). Using these models we can also examine Child × Instruction interactions. We used these models to follow 86 children (including Aaron) from 40 first-grade classrooms into 33 second-grade classrooms. Classrooms were observed in the fall, winter, and spring, and children's vocabulary and word-reading skills were assessed fall and spring. Again, using the dimensions of instruction, we then assigned the subactivities to a quadrant (see Table 10.2), teacher managed versus child managed and code focused versus meaning focused for each grade. Thus, four instructional

TABLE 10.2
Interactions Between Instructional Dimensions and Child Characteristics

Instruction dimension	Child characteristics	
	High	Low
	Decoding	
Teacher managed, code focused	No change	Increase
	Vocabulary	
Child managed, meaning focused	Increase	Decrease

variables were identified for each grade (1 and 2), and each variable represented the amount of instruction provided in minutes per day.

On average, students exhibited gains in vocabulary and word-reading skills from the beginning of first to the end of second grade, but there was notable variability. For example, from the beginning of first to the end of second grade, students' word-reading skills ranged from beginning of kindergarten to high school grade equivalent levels. The results of the cross-classified random effects model revealed that both first- and second-grade teacher-managed, code-focused instruction related to students' word-reading skill growth. However, the direction and magnitude of the effect depended on students' fall first-grade word-reading score. Generally, students who were in classrooms with higher amounts of teacher-managed, code-focused instruction in both first and second grade demonstrated greater letter- and word-reading skill growth by the end of second grade than did students in first- and second-grade classrooms with less teacher-managed, code-focused instruction (i.e., higher spring scores controlling for initial letter and word reading and vocabulary). However, as we saw before, in first grade, students with lower initial letter- and word-reading scores demonstrated greater letter- and word-reading skill growth in classrooms with higher amounts of teacher-managed, code-focused instruction (see Figure 10.1; Connor, Morrison, & Underwood, 2007). Students with stronger letter- and word-reading skills at the beginning of first grade demonstrated less letter- and word-reading growth in these same classrooms. First-grade teacher-managed, code-focused instruction had little effect on scores for students with strong letter- and word-reading scores at the beginning of first grade (see Figure 10.2). In second grade, there was a positive effect for teacher-managed, code-focused instruction. However, the effect was greater for students who started first grade with stronger word reading skills than it was for students with weaker skills. Although counterintuitive, plotted out, the magnitude of the cumulative effect of first and second grade is compelling (see Figures 10.1 and 10.2).

So what does this mean for Aaron? Remember that he had reading skills below grade expectations at the beginning of first grade, and although he got

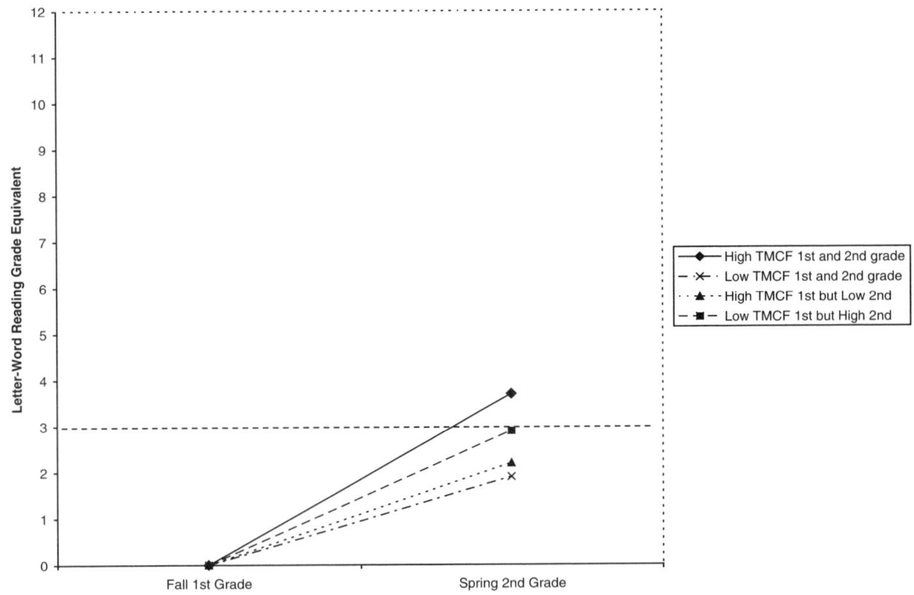

Figure 10.1. Modeled word-reading growth for children who received different patterns of teacher-managed, code-focused (TMCF) instruction in first and second grade and who began first grade with vocabulary and reading skills falling well below grade and age expectations. Dashed line denotes third-grade level achievement expected by the end of second grade.

above average amounts of teacher-managed, code-focused instruction, he did not receive enough to bring him to grade level by the end of first grade. However, second grade offered a second chance—if he received enough teacher-managed, code-focused in second grade, he would theoretically have been able to achieve grade level skills by the end of second grade. Without teacher-managed, code-focused instruction in second grade, the models predicted that he would not demonstrate growth in word-reading skills. As expected, Aaron received less than 1 minute a day of teacher-managed, code-focused instruction, and his word-reading skills stagnated. He showed no growth in word-reading skills in second grade and consequently, by the end of second grade, he was a full grade and a half behind his peers.

CONSEQUENCES OF INSTRUCTIONAL DISCONTINUITY

As these examples from our research and others illustrate, the schooling experiences of American children can be highly fragmented with enormous consequences for their academic progress. Three salient features of this piecemeal approach to children's development deserve consideration, along with their potential consequences.

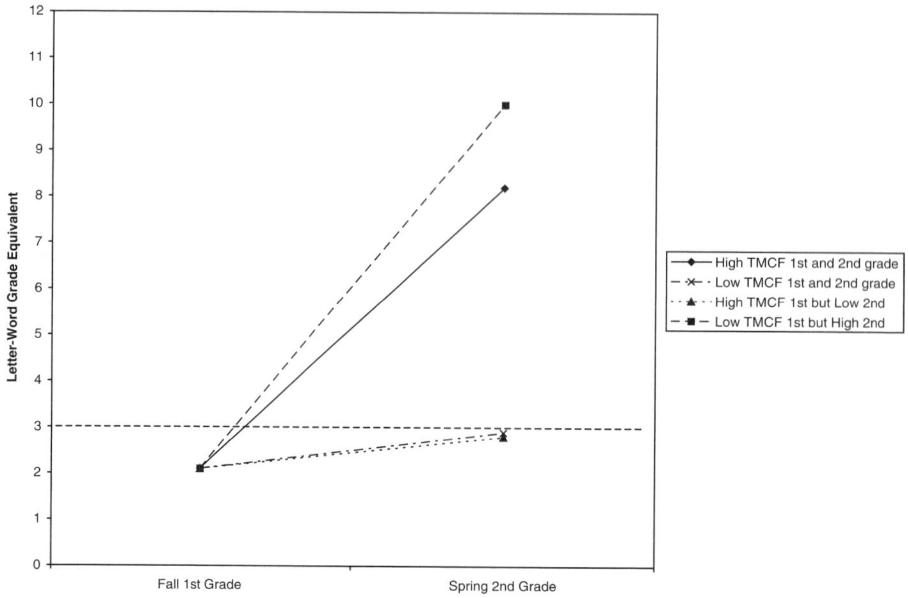

Figure 10.2. Modeled word-reading growth for children who received different patterns of teacher-managed, code-focused (TMCF) instruction in first and second grade and who began first grade with vocabulary and reading skills falling well above grade and age expectations. Dashed line denotes third-grade level achievement expected by the end of second grade.

Variability of Environmental Experiences Over Time

The instructional discontinuity across grades, inherent in our educational system, has magnified the variability of educational experiences that children receive in our society. As a consequence, substantial numbers of individual children are not receiving the amounts and types of instruction they need to grow optimally in each grade. For increasing numbers of children each year, these practices produce an ongoing stream of mismatches over time between what children would most benefit from and what they actually receive. Hence, sizable numbers of children with relatively strong academic profiles become bored and fail to progress, and a significant percentage of children with weaker skills become frustrated and fall further behind. The long-term impact is the creation of cumulatively growing numbers of children disengaged from school and its mode of instruction.

Ignoring Development

In a very direct way, the discontinuity of educational experiences offered to children ignores the continuity of development that the transactional view emphasizes. Child development involves the continuous dynamic interaction over time between the child and the environmental context. With parent–child interaction there is for the most part relative continuity in the nature of the interactional patterns and therefore the experiences the child receives. Our data have revealed this is not so with schooling. In part, the problem in the educational community is the simple lack of knowledge on the part of teachers in subsequent grades of the skills and other characteristics of the children they are about to teach. It is not customary for teachers to have extensive knowledge of the skill levels of the children in their incoming classes, nor is systematic sharing of assessments or even clinical impressions by teachers in different grades common practice, No Child Left Behind notwithstanding. The situation is exacerbated even further when one crosses the boundary between preschool and elementary school, a transition increasingly considered a crucial academic hurdle. Overall, the transactional discontinuity characteristic of schooling experiences actively ignores the importance of continuous ongoing assessment of children, sharing knowledge across teachers and grades and using knowledge of the child's skill levels to program effective instruction.

Abandoning Individualization

In practice, the disconnected movement of children from grade to grade and the general reform practice of highly scripted core curriculum with every child on the same page means that too often teachers, in planning their instruction, focus not on individual children and their learning but rather on curriculum and classrooms. Instruction is pitched at a particular segment of the class depending on the teacher's overall intuition about the average or modal level of students' functioning. In our first-grade study, the reading curriculum emphasized a whole-language approach, which clearly was geared to and most benefited the strongest readers. As a consequence, the weaker readers did not receive the amounts and types of instruction most effective for them, and they fell further behind. There are recent reports that the reverse is happening at present; in the pressure to help disadvantaged children and to show adequate yearly progress, school districts are mandating large chunks of time on basic skills instruction with the consequence that some stronger students are not progressing as expected.

TAKING THE TRANSACTIONAL VIEW SERIOUSLY

Within the environmental context of schooling, it is clear from the results we have presented that individualizing instruction is a vital step in our efforts to improve literacy in America. To achieve this worthy goal we will need to take the notion of development seriously and to embrace the transactional nature of developmental processes. Early and ongoing assessment of students' language and literacy skills, coupled with information sharing across grades and between teachers and parents, will facilitate the focus on individual children and their learning and help to provide the continuity of experiences needed to maximize their individual growth.

REFERENCES

Anderson, R. C., & Freebody, P. (1981). Vocabulary knowledge. In J. T. Guthrie (Ed.), *Comprehension and teaching: Research reviews* (pp. 77–111). Newark, DE: International Reading Association.

Beals, D. E., & DeTemple, J. M. (1993). Home contributions to early language and literacy development. *Yearbook of the National Reading Conference, 42*, 207–215.

Beck, I. L., McKeown, M. G., & Kucan, L. (2005). Choosing words to teach In E. H. Hiebert & M. L. Kamil (Eds.), *Teaching and learning vocabulary: Bringing research to practice* (pp. 209–222). Mahwah, NJ: Erlbaum.

Cameron, C. E., Connor, C. M., Morrison, F. J., & Jewkes, A. M. (2008). Effects of classroom organization on letter–word reading in first grade. *Journal of School Psychology, 46*, 173–192).

Cameron, C. E., McClelland, M. M., Jewkes, A. M., Connor, C. M., Farris, C. L., & Morrison, F. J. (2008). Touch your toes! Developing a behavioral measure of preschool self-regulation. *Early Childhood Research Quarterly, 23*, 141–158.

Connor, C. M., Jakobsons, L. J., Crowe, E., & Granger, J. (in press). Instruction, differentiation, and student engagement in Reading First classrooms. *Elementary School Journal*.

Connor, C. M., Morrison, F. J., Fishman, B. J., Schatschneider, C., & Underwood, P. (2007, January). The early years: Algorithm-guided individualized reading instruction. *Science, 315*, 464–465.

Connor, C. M., Morrison, F. J., & Katch, E. L. (2004). Beyond the reading wars: Exploring the effect child–instruction interactions on growth in early reading. *Scientific Studies of Reading, 8*, 305–336.

Connor, C. M., Morrison, F. J., & Petrella, J. N. (2004). Effective reading comprehension instruction: Examining Child × Instruction interactions. *Journal of Educational Psychology, 96*, 682–698.

Connor, C. M., Morrison, F. J., & Slominski, L. (2006). Preschool instruction and children's literacy skill growth. *Journal of Educational Psychology, 98*, 665–689.

Connor, C. M., Morrison, F. J., & Underwood, P. S. (2007). A second chance in second grade: The independent and cumulative impact of first- and second-grade reading instruction and students' letter-word reading skill growth. *Scientific Studies of Reading, 11*, 199–233.

Cronbach, L. J., & Snow, R. E. (1977). *Aptitudes and instructional methods: A handbook for research on interactions*. New York: Irvington.

Cunningham, P., & Hall, D. (1998). The four blocks: A balanced framework for literacy in primary classrooms. In K. R. Harris, S. Graham, & D. Deshler (Eds.), *Teaching every child every day: Learning in diverse schools and classrooms* (pp. 32–76). Cambridge, MA: Brookline Books.

Dahl, K. L., & Freppon, P. A. (1995). A comparison of innercity children's interpretations of reading and writing instruction in the early grades in skills-based and whole language classrooms. *Reading Research Quarterly, 30,* 50–74.

Dickinson, D. K., Anastasopolous, L., McCabe, A., Peisner-Feinberg, E. S., & Poe, M. D. (2003). The comprehensive language approach to early literacy: The interrelationships among vocabulary, phonological sensitivity, and print knowledge among preschool-aged children. *Journal of Educational Psychology, 95,* 465–481.

Dickinson, D. K., & Tabors, P. O. (2001). *Beginning literacy with language.* Baltimore: Brookes Publishing.

Foorman, B. R., Francis, D. J., Fletcher, J. M., Schatschneider, C., & Mehta, P. (1998). The role of instruction in learning to read: Preventing reading failure in at risk children. *Journal of Educational Psychology, 90,* 37–55.

Foorman, B. R., Schatschneider, C., Eakin, M. N., Fletcher, J. M., Moates, L. C., & Francis, D. J. (2006). The impact of instructional practices in Grades 1 and 2 on reading and spelling achievement in high poverty schools. *Contemporary Educational Psychology, 31,* 1–29.

Goodman, K. (1970). Reading: A psycholinguistic guessing game. In H. Singer & R. B. Ruddell (Eds.), *Theoretical models and processes of reading* (pp. 259–272). Newark, DE: International Reading Association.

Guthrie, J. T., Schafer, W. D., & Huang, C.-W. (2001). Benefits of opportunity to read and balanced instruction on the NAEP. *Journal of Educational Research, 94,* 145–162.

Hart, B., & Risley, T. R. (1995). *Meaningful differences in the everyday experience of young American children.* Baltimore: Brookes Publishing.

Hoover, W. A., & Gough, P. B. (1990). The simple view of reading. *Reading and Writing, 2,* 127–160.

Juel, C., & Minden-Cupp, C. (2000). Learning to read words: Linguistic units and instructional strategies. *Reading Research Quarterly, 35,* 458–492.

Loban, W. (1976). *Language development: Kindergarten through grade twelve.* Urbana, IL: National Council of Teachers of English.

Mason, J. M., & Stewart, J. P. (1990). Emergent literacy assessment for instructional use in kindergarten. In L. M. Morrow & J. K. Smith (Eds.), *Assessment for instruction in early literacy* (pp. 155–175). Englewood Cliffs, NJ: Prentice-Hall.

McCardle, P., & Chhabra, V. (Eds.). (2004). *The voice of evidence in reading research.* Baltimore: Brookes Publishing.

McClelland, M. M., Cameron, C. E., Connor, C. M., Farris, C. L., Jewkes, A. M., & Morrison, F. J. (2007). Links between behavioral regulation and preschoolers' literacy, vocabulary, and math skills. *Developmental Psychology, 43,* 947–459.

Morrison, F. J., Bachman, H. J., & Connor, C. M. (2005). *Improving literacy in America: Guidelines from research*. New Haven, CT: Yale University Press.

Morrison, F. J., Smith, L., & Dow-Ehrensberger, M. (1995). Education and cognitive development: A natural experiment. *Developmental Psychology, 31*, 789–799.

National Institute of Child Health and Development Early Child Care Research Network. (2005). Pathways to reading: The role of oral language in the transition to reading. *Developmental Psychology, 41*, 428–442.

National Reading Panel. (2000). *National Reading Panel report: Teaching children to read: An evidence-based assessment of the scientific research literature on reading and its implications for reading instruction* (NIH Publication No. 00-4769). Washington, DC: U.S. Department of Health and Human Services, Public Health Service, National Institutes of Health, National Institute of Child Health and Human Development.

Poe, M. D., Burchinal, M. R., & Roberts, J. E. (2004). Early language and the development of children's reading skills. *Journal of School Psychology, 42*, 315–332.

Pressley, M. (1998). *Reading instruction that works: The case for balanced teaching*. New York: Guilford Press.

Raudenbush, S. W., & Bryk, A. S. (2002). *Hierarchical linear models: Applications and data analysis methods* (2nd ed.). Thousand Oaks, CA: Sage.

Ravitch, D. (2001). It is time to stop the war. In T. Loveless (Ed.), *The great curriculum debate: How should we teach reading and math?* (pp. 210–228). Washington DC: Brookings Institution Press.

Rayner, K., Foorman, B. R., Perfetti, C. A., Pesetsky, D., & Seidenberg, M. S. (2001). How psychological science informs the teaching of reading. *Psychological Science in the Public Interest, 2*(2), 31–74.

Ross, S. M., Smith, L. J., Slavin, R. E., & Madden, N. A. (1997). Improving the academic success of disadvantaged children: An examination of success for all. *Psychology in the Schools, 34*, 171–180.

Sameroff, A. J., & MacKenzie, M. J. (2003). Research strategies for capturing transactional models of development: The limits of the possible. *Development and Psychopathology, 15*, 613–640.

Sanders, W. L., & Horn, S. P. (1998). Research findings from the Tennessee value-added assessment system (TVAAS) database: Implications for educational evaluation and research. *Journal of Personnel Evaluation in Education, 12*, 247–256.

Scarborough, H. S. (1990). Very early language deficits in dyslexic children. *Child Development, 61*, 1728–1743.

Scarborough, H. S. (1998). Early identification of children at risk for reading disabilities: Phonological awareness and some other promising predictors. In B. K. Shapiro, P. J. Accardo, & A. J. Capute (Eds.), *Specific reading disability: A view of the spectrum* (pp. 75–119). Timonium, MD: York Press.

Scarborough, H. S. (2001). Connecting early language and literacy to later reading (dis)abilities: Evidence, theory, and practice. In S. B. Neuman & D. K. Dickinson (Eds.), *Handbook of early literacy research* (pp. 97–110). New York: Guilford Press.

Schatschneider, C. (2004). *A multivariate study of individual differences in performance on the reading portion of the Florida Comprehensive Assessment Test: A brief report.* Retrieved August 4, 2006, from http://www.fcrr.org/TechnicalReports/Multi_variate_study_december2004.pdf

Senechal, M., & LeFevre, J.-A. (2001). On refining theoretical models of emergent literacy: The role of empirical evidence. *Journal of School Psychology, 39,* 439–460.

Senechal, M., & LeFevre, J.-A. (2002). Parental involvement in the development of children's reading skill: A five-year longitudinal study. *Child Development, 73,* 445–460.

Senechal, M., LeFevre, J.-A., Thomas, E. M., & Daley, K. E. (1998). Differential effects of home literacy experiences on the development of oral and written language. *Reading Research Quarterly, 33,* 96–116.

Snow, C. E. (2002). *Reading for understanding: Toward an R & D program in reading comprehension.* Arlington, VA: RAND.

Snow, C. E., Burns, M. S., & Griffin, P. (Eds.). (1998). *Preventing reading difficulties in young children.* Washington, DC: National Academies Press.

Spira, E. G., Bracken, S. S., & Fischel, J. E. (2005). Predicting improvement after first-grade reading difficulties: The effects of oral language, emergent literacy, and behavior skills. *Developmental Psychology, 41,* 225–234.

Storch, S. A., & Whitehurst, G. J. (2002). Oral language and code-related precursors to reading: Evidence from a longitudinal structural model. *Developmental Psychology, 38,* 934–947.

Taylor, B. M., Pearson, D. P., Clark, K., & Walpole, S. (2000). Effective schools and accomplished teachers: lessons about primary-grade reading instruction in low-income schools. *Elementary School Journal, 101,* 121–165.

Torgesen, J. K., Burgess, S., Wagner, R. K., & Rashotte, C. A. (1994). Longitudinal studies of phonological processing and reading. *Journal of Learning Disabilities, 27,* 276–286.

Torgesen, J. K., Wagner, R. K., Rashotte, C. A., Rose, E., Lindamood, P., Conway, T., & Garvan, C. (1999). Preventing reading failure in young children with phonological processing disabilities: Group and individual responses to instruction. *Journal of Educational Psychology, 91,* 579–593.

Whitehurst, G. J., & Lonigan, C. J. (1998). Child development and emergent literacy. *Child Development, 69,* 335–357.

11

PARENT LEARNING SUPPORT AND CHILD READING ABILITY: A CROSS-LAGGED PANEL ANALYSIS FOR DEVELOPMENTAL TRANSACTIONS

ELIZABETH T. GERSHOFF, J. LAWRENCE ABER, AND MARGARET CLEMENTS

The past half century of developmental psychology has moved away from a unidirectional understanding of development (e.g., parents affect children and not vice versa) and toward a bidirectional conceptualization of development. Children are now assumed to affect (Bell, 1968), and even select (Scarr & McCartney, 1983), their environments inasmuch as their environments affect children's behavior. Indeed, key among many of the most influential developmental theories in the past several decades is the assumption that children have bidirectional, or reciprocal, relationships with their environments (e.g., Bandura, 1978; Bronfenbrenner, 1986; Lerner, 1991).

Sameroff and Chandler (1975) took the idea of reciprocal relations one step further by arguing that children and contexts influence each other mutually over time, such that the change a child precipitates in the environment will change how the environment in turn affects the child at a later time point. The transactional model that they proposed held that children influence their environments to an extent that is commensurate with the influence the environments have on the children (Sameroff & MacKenzie, 2003).

This research was supported by Grant 5R01HD042144 from the National Institute of Child Health and Human Development awarded to Elizabeth T. Gershoff and J. Lawrence Aber.

The developmental context in which transactions are most often studied has been that of parents and their parenting behavior toward their children. When parents and children influence each other in a mutually positive way, it is said that the parent–child relationship is characterized by reciprocity (Maccoby & Martin, 1983). Transactional influences can also be negative, such as when a parent's harsh behavior increases a child's difficult behavior, which then feeds back into parent harsh behavior, and so on, in a coercive cycle (Patterson, 1982).

Because the developmental and social sciences are concerned with many phenomena that do not readily lend themselves to "gold standard" randomized-control experimental designs, inferring causality in research on children and families can be difficult. Cross-lagged panel analysis (CLPA) continues to be a crucial tool for researchers to detect transactional processes between parent and child behavior over time. Originally developed as a counterpoint to cross-sectional studies of relations among variables, CLPA can assess the direction of influence between variables when both are measured across two or more waves of data through cross-lagged paths (Lazarsfeld, 1940; Lazarsfeld & Fiske, 1938). By allowing a given independent variable to precede a dependent variable in time, CLPA satisfies one necessary condition for determining a causal effect of one variable on another (Davis, 1985). When CLPA was initially proposed as a statistical method for identifying and ruling out potential causal pathways, it involved the comparison of multiple bivariate correlations, typically across two waves of data (see Kenny, 1975). However, because this approach relies on bivariate correlations, each relation between two variables of interest is obtained in isolation from all other possible relations with other variables in the model. The advent of structural equation modeling (SEM) allowed the simultaneous estimation of all paths in the CLPA model to control for all potential relations among the variables in the calculation of any given path (Martens & Haase, 2006).

REVEALING CAUSAL PROCESSES WITH CROSS-LAGGED PANEL ANALYSIS

Most applications of CLPA in psychology have used a discrete-time form of the analysis in which the within-time relations between two variables are modeled as correlations rather than as directional paths (e.g., Burt, McGue, & Iacono, 2006; Halonen, Aunola, & Ahonen, 2006; Rueter & Conger, 1998). Such models are called *discrete-time CLPA* because the causal effects of a parent on a child, for example, are modeled longitudinally through a cross-lagged path across a specific (i.e., discrete) time interval. Such longitudinal paths allow more confident assumptions about causality. In SEM, correlations among variables are not included in the equations that predict dependent variables and thus do not contribute to the explanation of variance in endogenous variables (Kline, 2004). As a result, the correlations between the variables' error terms within a given time point can help improve the fit of a model but are not being used to predict the dependent variables.

The assumption at the heart of discrete-time CLPA, namely that causality is only demonstrated across time intervals and does not involve causal effects that occur within a given time window, does not fit with our current theories of child development and of parent–child relationships. Theories that development is dominated by discrete, qualitative changes, such as that of Piaget (1928/1959), have now been largely replaced by appreciation of development as a more gradual and additive process (e.g., Rogoff, 1998). In the realm of parent–child relationships, the effect of a child's behavior at $time_t$ on his or her parent's behavior at $time_{t+1}$ is but one chance observation of how the child affects the parent, sampled from countless interactions across the course of their relationship. Although researchers must use discrete observations of parent and child behavior, the individuals studied have thousands if not millions of opportunities to interact and influence each other's behavior in the intervals between observations (Oud, 2002). We argue that an important implication of viewing relationship partners as mutually influencing each other over time is that small effects of one partner's behavior on another will accumulate over time. In other words, a child's behavior may lead to small immediate adjustments in the parent's behavior, and these small adjustments may add up to marked differences in parent behavior over months or years.

Discrete-time CLPA thus does not fully capture the process by which a child's behavior changes the behavior of a parent and vice versa. In other words, what does theory predict are the reasons how or why a child's behavior should affect the parent's behavior in the next observation, which may be days, months, or even years later? To illustrate the need to consider process, we use as illustration the hypothesis that a child's aggression at one time point will elicit harsh discipline from his or her parent at a later point (Gershoff, 2002). This hypothesis can be tested by analyzing a cross-lagged model that uses children's rate of hitting peers at age 3 to predict their parents' harsh discipline at age 5 and interpret a significant cross-lagged path coefficient as a causal child effect (for a similar example, see Kandel & Wu, 1995). However, we must consider the situation more carefully from a process standpoint. Exactly how would a child's rate of behavior at one time point have an independent effect on a parent's disciplinary behavior 2 years later? Do we believe that parents are so slow to adapt their parenting in reaction to their child's behavior that it takes them 2 years to do so? Of course not, but this is precisely what the cross-lagged path from child to parent is testing in the discrete-time CLPA when the time points are separated by long intervals. The implied process is that the effect of a child on a parent has a considerable time delay.

If we reject such a process as far-fetched, we can try another. Is the child's behavior so powerful an influence that it continues to exert a predictive influence on parent behavior 2 years later, regardless of intervening changes in the child's behavior? Said another way, would we really expect parents' style of disciplining their 5-year-old to be predicted more by the child's aggression 2 years earlier than the child's current aggression at age 5? To the extent that the child's aggression at age 3 represents an aspect of his or her temperament, it is somewhat reasonable to expect that age 3 aggression may predict future received parenting. However, we propose that this lagged effect is better understood in the context of children's current behavior and its effect on current parent behavior.

There is an alternative to discrete-time CLPA that better captures the processes involved in transactional relationships. *Continuous-time CLPA* incorporates the potential to examine how partners simultaneously influence each other within any given observation window, thereby avoiding the oversimplification of family processes as occurring only across large gaps in time with nothing happening in between (Oud, 2002). To capture potential reciprocal effects, or feedback loops, continuous-time CLPA models include two directional paths between the measured variables within a given time point with parent and child variables simultaneously serving as predictor and outcome variables (Oud, 2002). Figure 11.1 displays a continuous-time CLPA, in which there are two sets of directional, within-time paths from parent to child (f_2 and f_3) and from child to parent (g_2 and g_3) rather than within-time correlations as in traditional discrete-time CLPA. These parallel directional paths are for exogenous (downstream) variables only (i.e., variables at $time_2$ and $time_3$) and constitute within-time reciprocal or bidirectional effects. The correlational path (c_1) between parent and child variables is only retained at $time_1$. Although discrete-time CLPA models may satisfy methodological

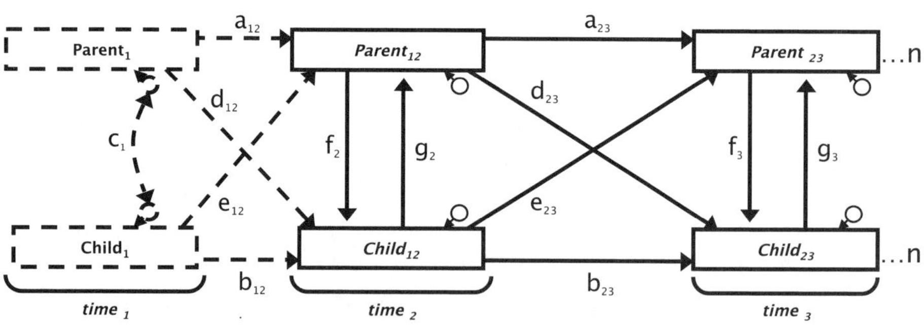

Figure 11.1. Continuous-time cross-lagged panel analysis with a fixed effects approach. Subscripts with one number indicate within-time paths; subscripts with two numbers indicate across-time paths (e.g., a_{12} links the parent variables at $time_1$ and $time_2$). Dashed paths reflect fixed effects.

requirements for inferring causality because predictors always precede criterion variables (Cole & Maxwell, 2003; Gollob & Reichardt, 1987), we argue that the assumption that the events are "discrete" events with a beginning and an end is not always the best or most accurate representation of the transactional nature of human relationships in particular (although the model applies to other variables as well). As the name implies, continuous-time CLPA presumes that the phenomena under study are unfolding continually over time, sometimes in a large number of microlevel transactions. For the conveniences of research, we may measure a particular parenting construct at one point in time, but we do so with the understanding that that particular measurement occasion is but a snapshot of the true nature of the parent behavior that persists over time.

Despite increasing theoretical emphasis on transactional processes and reciprocity of influence between children and their environments, CLPA models that incorporate reciprocal paths are only rarely used in developmental psychology. A practical reason why is the fact that such models with reciprocal paths, also called *nonrecursive models*, are more difficult to fit, primarily because they are easily underidentified (i.e., they have fewer observed variables than there are model parameters) and thus insoluble (Kline, 2004). However, several statistical solutions to this problem have been offered over the years (for a discussion of such methods, see Martens & Haase, 2006), paving the way for greater use of nonrecursive models, including continuous-time CLPA.

Another reason to use continuous-time CLPA is that the magnitude and even the sign of their cross-lagged coefficients in discrete-time CLPA models have been found to be heavily dependent on the length of the time interval chosen (Oud, 2002). Delsing, Oud, and De Bruyn (2005) demonstrated that cross-lagged effects in a discrete-time cross-lagged model of the effects of family relationships on adolescent problem behavior were substantially different when the time lag was 1 year versus 2 years in length. In contrast, the coefficients in their continuous-time cross-lagged model were robust to the length of time between observations. Thus, continuous-time CLPA models are likely to obtain more robust, stable estimations of path coefficients than discrete-time CLPA.

CROSS-LAGGED PANEL ANALYSIS AND FIXED EFFECTS

Because both parents and children develop over time (Sigel & Parke, 1987), it is not enough to say that a measure of parenting at one point predicts change in a child's behavior from $time_1$ to $time_2$ or vice versa. Rather, a preferred measure of the dynamic nature of the parent–child relationship is whether a change in one partner's behavior predicts a change in the other partner's behavior. Indeed, the associations of change in one variable with changes in another variable can reveal patterns of causality. For example, if a parent is presumed to have a causal effect on a child, then when the parent's behavior changes, so too should the child's behavior change. If the child's behavior does not change, then the parent cannot be said to have a causal effect on the child (Kessler & Greenberg, 1981), at least over that specified time period. The processes by which change in one individual or setting in turn changes the behavior of another individual or setting are what is meant by *transactions* (Sameroff & MacKenzie, 2003).

A focus on intraindividual change is characteristic of what are called *fixed effects* models (Duncan, Magnuson, & Ludwig, 2004). Specifically, when one wave of a variable is regressed on (i.e., predicted by) an earlier measurement of the same variable, the residual variance for the second variable can be used to predict and/or be predicted by other variables. In CLPA, the autoregressive paths between repeated measures of a variable are typically considered *controlled for* and discounted as not interesting in developmental models because they are measures of stability, whereas developmentalists are most interested in change. However, when viewed as part of a fixed effects model, these repeated measures are no longer indexes of stability and are instead key measures of change over time. Once the portion of variance in the second measure that is shared with the first measure is accounted for (its stability across autoregressive paths), the variance that remains is change from the earlier wave. Specifically, the autoregressive path is an index of stability, whereas the residual variance is an index of change. This residual variance or change can then be used to predict and/or be predicted by other variables in a model. Of course, this residual change score can be calculated for other repeated measures in a model, resulting in multiple such change scores for various measures. When multiple residual change scores are created in a model, as in CLPA, and when they are regressed on one another, this regression of "change on change" represents a more adequate index of a transactional effect.

A fixed-effects approach to a continuous-time CLPA requires that the first wave not be interpreted because it is used solely to produce the first change coefficient (e.g., from $time_1$ to $time_2$). We thus depict paths within or from $time_1$ in Figure 11.1 with dashed lines. The main path coefficients to be interpreted in such a continuous-time CLPA are only those within and between $time_2$ and $time_3$. The labels $parent_{12}$, $child_{12}$, $parent_{23}$, and $child_{23}$ indicate that these variables are considered to be residual change scores including the two data points in the subscript. In this model, reciprocal paths do not represent within-time cross-sectional paths but rather the influence of change in the parent over the course of a time interval (such as from the beginning of kindergarten to the end of first grade) on the change in the child over the same time period and vice versa.

This approach shares some characteristics of the latent difference score (LDS) model (Ferrer & McArdle, 2003) in which the differences between measures of the same variable at subsequent waves are considered to represent dynamic change. The main difference between the model we propose and the LDS model is that in the latter the difference between repeated measures of a variable is modeled as a latent construct, whereas our model considers the downstream autoregressive effects to be measures of residual change.

INDIRECT EFFECTS AND TRANSACTIONAL PROCESSES

We argue that any effect of the child's behavior at $time_1$ on the parent's behavior at $time_2$ is not apt to be a direct one, as would be implied by a significant cross-lagged path, but rather an indirect, or mediated, one. Specifically, our interpretation of continuous-time CLPA is that within-time reciprocal paths are integral to understanding causal relations over time. In contrast to the typical assumptions made in CLPA, indirect effects are a way to explicitly model the processes by which a child's behavior at a given time point affects his or her parent's behavior in the future and vice versa. Structural equation models with one or more mediator variables allow the decomposition of a total effect of one variable on another variable into direct and indirect components (Kessler & Greenberg, 1981; Kline, 2004).

Modeling indirect or mediating effects in SEM is straightforward. Although most psychologists are familiar with the Baron and Kenny (1986) three-regression method of detecting mediation, in SEM the direct and indirect (mediated) relations among variables can be estimated simultaneously in one model. Software packages used to conduct SEM can easily calculate direct, indirect, and total effects (e.g., Amos: Arbuckle & Wothke, 1995; Mplus: Muthén & Muthén, 2007), and guides for how to calculate the significance of mediated effects are also available (see MacKinnon & Dwyer, 1993).

In a continuous-time CLPA, there are three pathways through which parent behavior at $time_t$ can affect child behavior at $time_{t+1}$, (and, conversely, how child behavior at $time_t$ can affect parent behavior at $time_{t+1}$). The cross-lagged paths between them (see paths d_{23} and e_{23} in Figure 11.1) constitute direct effects of one on the other at a later time point. For the prediction of $child_{23}$ from $parent_{12}$ there are two indirect pathways (see circle-dotted paths in Figure 11.2). The first is mediated through $parent_{23}$, and the second is mediated through $child_{12}$. There are also two indirect effects of $child_{12}$ on $parent_{23}$. One path is through $parent_{12}$ and another is through $child_{23}$ (see dotted paths in Figure 11.2). What is clear from these indirect pathways is that they each include one of the autoregressive/stability pathways. As we noted previously, it is difficult to imagine how a child's behavior at a given $time_1$ could have a direct effect on a parent's behavior at another time point 1 or 2 years later without taking into account the stability of the child's behavior across time. To illustrate our recommended approach to transactional analysis—one based

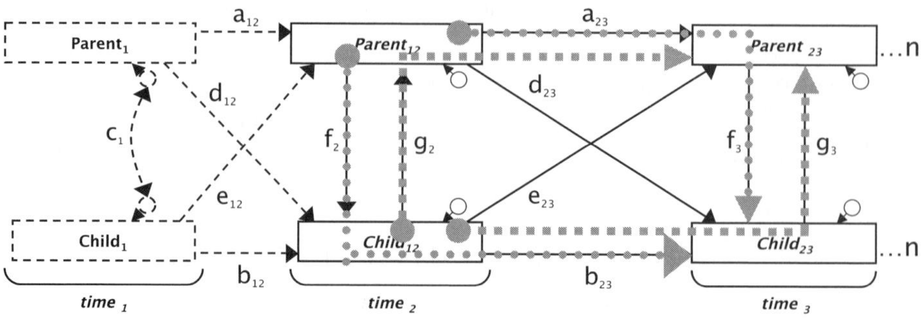

Figure 11.2. Indirect effects of Δ Parent_{12} on Δ Child_{23} (circle-dotted paths) and Δ Child_{12} on Δ Parent_{23} (square-dotted paths). Dashed paths reflect fixed effects.

on continuous-time CLPA that includes both fixed effects and indirect effects components—we consider potential reciprocal relations among parents' investment in their children and their children's academic achievement.

PARENT LEARNING SUPPORT AND CHILD READING ABILITY ACROSS THREE WAVES

One economic perspective on parent–child relations argues that the amount of money parents spend on their children (such as by purchasing books, toys, or high-quality child care) and the time they spend with them in joint activities (including visiting the library or doing a science project at home) are considered investments that have the potential to enhance children's cognitive competencies (Foster, 2002; Haveman & Wolfe, 1994). This perspective is easily recognizable as a unidirectional model: Parents make the decisions about how to allocate money, time, and energy, and children are the beneficiaries (or not) of their decisions.

Although several studies have confirmed that parents' investments in their children do in fact predict higher levels of achievement and/or lower levels of behavior problems (e.g., Gershoff, Aber, Raver, & Lennon, 2007; Guo & Harris, 2000; Linver, Brooks-Gunn, & Kohen, 2002), none of these studies have viewed the intrafamilial investment process as transactional, nor have they accounted for potential reciprocal effects of children's achievement on parents' investments. Transactional theory (Sameroff & Chandler, 1975) would predict that children should exert influence on their parents in addition to their parents exerting influence on them and thus would predict that children's levels of academic achievement should elicit differential parent investments as parent investments predict children's achievement. To simplify our discussion of the components of the continuous-time CLPA model, we focus in this example on one aspect of children's achievement, namely their reading ability from kindergarten to third grade, and on one type of parent investment most likely to directly impact child reading, parent's learning support of their children over the early school years. (Because of our simplifying assumption, this analysis is vulnerable to omitted-variable bias. We will be testing a more complete version of this hypothesis in a later work.)

We can imagine several ways that parent learning support and child reading ability may interact in transactional processes. In the case of children's reading ability, children who express an early interest in reading may inspire their parents to read to them and help them practice their own reading. Positive reading experiences with parents will increase the children's own reading ability, thus completing a transactional cycle. Yet it is also possible that children who have early difficulty with reading may make it more likely that their parents will spend more time helping them with their reading, which we hope would help improve children's reading abilities in the long run. In this latter case, the first pathway would be negative (less reading ability associated with more parent reading to children), whereas the second would be positive (more parent reading to children associated with higher child reading ability).

We set out to examine the extent to which parent learning support and children's reading ability are mutually influential using the continuous-time CLPA model described previously. We used three waves of data from 16,919 families involved in the Early Childhood Longitudinal Study, Kindergarten Class of 1998–1999 (ECLS-K; for complete descriptions of this study, see Gershoff et al., 2007; Rathbun & West, 2004). In each of the kindergarten, first-grade, and third-grade waves, parents reported on their learning support behaviors (through the cognitive stimulation scale of the HOME Inventory; Caldwell & Bradley, 1984). Also at each wave, children's reading ability was directly assessed through standardized measures whose items were developed by the ECLS-K assessment work group (see Rathbun & West, 2004). For these analyses, we used item-response-theory-scored scales of reading achievement. Scores created in this way reflect the overall pattern of right and wrong answers to estimate ability and allow for the assessment of achievement gains in longitudinal analyses. In addition, the three models we tested controlled for parents' highest level of education and children's age at kindergarten entry. All analyses were conducted using the statistical software package Mplus Version 4.21 (Muthén & Muthén, 2007), including its Model Indirect function, which calculates the indirect and total effects for specified mediational pathways. We also used full-information maximum likelihood estimation in the presence of missing data, the current standard in the field (Schafer & Graham, 2002).

We acknowledge at the outset that our specified model is not ideal because the intervals between assessments are not equal. Several authors have argued that effect sizes are in part a function of the time lag between variables in a longitudinal design (Cole & Maxwell, 2003; Gollob & Reichardt, 1987).

Although we agree that equal time intervals are ideal, they are not always practical or possible. In this study, we decided the benefits derived from using a large, nationally representative sample outweighed the constraints of the study design available to us. Furthermore, the authors who have identified problems accruing to differential time lags have done so with discrete-time CLPAs. As noted previously, Delsing et al. (2005) reported that continuous-time CLPA models were more robust to differences in time lags than discrete-time CLPA models.

Even though ultimately we were interested in a model with cross-lagged and reciprocal paths, we compared the fit of three alternative models to determine which provided the best fit to the data. We began with the most basic model of panel data, a purely autoregressive model, and subsequently estimated models that increased in complexity, namely, the discrete-time cross-lagged model (see Figure 11.3) and the continuous-time cross-lagged model (see Figure 11.4). This allowed us to carefully confirm that the discrete-time cross-lagged model provided a better fit to the data than the autoregressive model (Martens & Haase, 2006), and then that the continuous-time model fit better than the discrete-time model (Delsing et al., 2005). Because the within-time correlations in the autoregressive and discrete-time CLPA models were not present in the continuous-time CLPA, these models were not nested within each other and thus could not be compared through chi-square difference tests. Rather, we compared several fit indexes to determine which model fit best, specifically which model or models met established criteria for the comparative fit index (CFI), root-mean-square error of approximation (RMSEA), and standardized root-mean-squared residual (SRMR; Hu & Bentler, 1999). In addition, we consulted the Akaike information criterion (AIC), which is a predictive fit index that can be used to select among nonnested, also known as *nonhierarchical*, models (Kline, 2004).

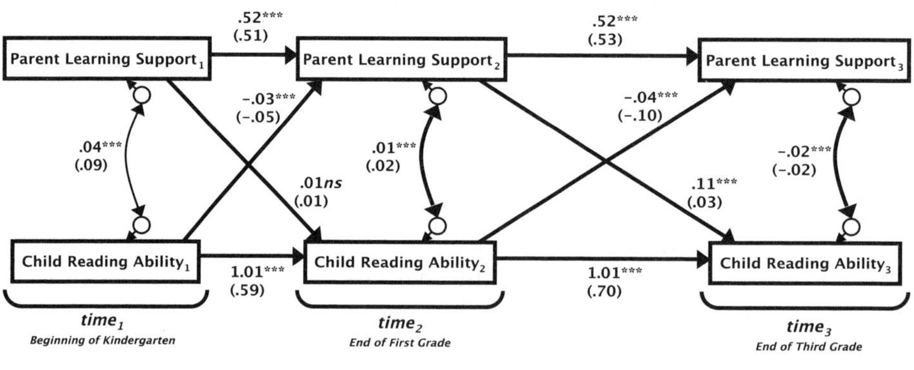

Figure 11.3. Parent learning support and child reading ability: discrete-time cross-lagged model. Comparative fit index = .956; root-mean-square error of approximation = .073; standardized root-mean-squared residual = .047; Akaike information criterion = 322,096. Standardized coefficients are presented in parentheses.

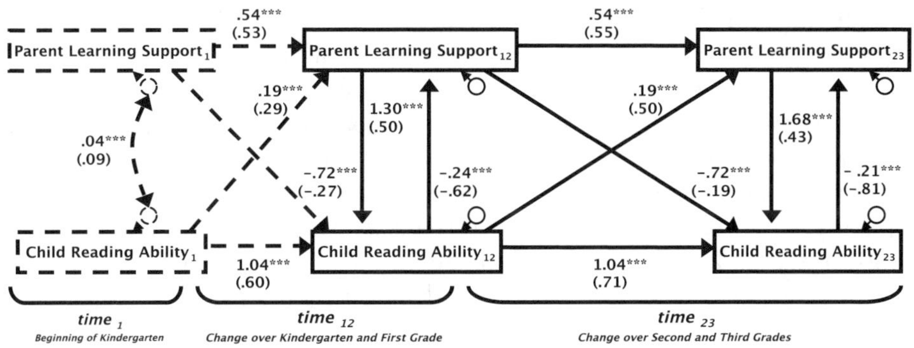

Figure 11.4. Parent learning support and child reading ability: continuous-time cross-lagged model. Comparative fit index = .953; root-mean-square error of approximation = .067; standardized root-mean-squared residual = .047; Akaike information criterion = 321,917. Standardized coefficients are presented in parentheses.

In the continuous-time CLPA model, we applied a fixed effects approach and thus considered the second and third measures of each construct to be residual change scores (paths involving time$_1$ are dashed in Figure 11.4). Thus, parent learning support$_{12}$ represents the extent to which parents on average change their level of learning support for their children over the course of their kindergarten and first-grade years (from beginning of kindergarten [time$_1$] to end of first grade [time$_2$]). Similarly, child reading ability$_{23}$ represents the change in children's reading ability over the 2 years of second and third grade (from end of first grade [time$_2$] to the end of third grade [time$_3$]). A major implication of our decision to interpret the residual variance in variables at time$_{t+1}$ (after controlling for variance in the same variable at time$_1$ by modeling these autoregressive paths) as change scores is that the reciprocal paths between the constructs portrayed in Figure 11.4 should be interpreted as the effects of change in parent learning support over 2 years on change in child reading ability over the same time period and vice versa, not as within-time effects of parent on child and child on parent.

As noted previously, a key problem with nonrecursive CLPA models is that specifying an identified model (a model for which a single solution cannot be obtained) can be difficult. Certain steps in specifying these models should be considered to ensure identification. For our continuous-time CLPA involving reciprocal paths, we imposed equality constraints as a means of achieving identification (see recommendations by Ferrer & McArdle, 2003; Kessler & Greenberg, 1981; Oud, 2002). Of other potential solutions, equality constraints are preferred because they involve relatively weak assumptions (Kessler & Greenberg, 1981). Accordingly, we imposed two types of equality constraints on longitudinal paths involving the same sequence of variables, resulting in paths that were held constant from the first to third time points.

One type of equality constraint we imposed involved within-variable autoregressive paths. For example, in Figure 11.4 the path parent learning support$_1$ → parent learning support$_{12}$ was set equal to parent learning support$_{12}$ → parent learning support$_{23}$. The second type of equality constraint involved the sets of cross-lagged paths in the model. For example, the path from parent learning support$_1$ → child reading ability$_{12}$ was set equal to parent learning support$_{12}$ → child reading ability$_{23}$. In the discrete-time CLPA model, these cross-lagged paths were freely estimated, and their coefficients were of the same magnitude (see Figure 11.3), and thus setting the parallel cross-lagged paths equal is reasonable.

All three models were at or near Hu and Bentler's (1999) recommendations for acceptable fit indexes for SEM analyses (CFI ≥ .950, RMSEA ≤ .060, and SRMR ≤ .080). The autoregressive model fit well, CFI = .950, RSMEA = .069, SRMR = .054, AIC = 322,284. However, the addition of cross-lagged paths in the discrete-time model improved the fit above that of the autoregressive model: CFI = .956, RMSEA = .073, SRMR = .047, AIC = 322,096, with a larger CFI, smaller SRMR, and smaller AIC, indicating that the discrete-time model is a better fitting one. In this discrete-time model, parent learning support predicted increased child reading ability, but significantly only for the time lag between the end of first grade and the end of third grade, $\beta_{12} = .01$ and $\beta_{23} = .03$ (see Figure 11.3), whereas higher child reading ability predicted lower levels of parent learning support across each of the lags, $\beta_{12} = -.05$ and $\beta_{23} = -.10$. In the text we present only standardized path coefficients; both standardized and unstandardized path coefficients are presented in all figures. The standardized coefficients for the cross-lagged paths in the discrete-time model are quite small, all at or below $|.10|$.

We then fit a continuous-time CLPA, which replaced the within-time covariances in the discrete-time model, with reciprocal directional paths between parent learning support and child reading ability within each wave (see Figure 11.4). This model provided the best fit to the data of the three models tested, CFI = .963, RMSEA = .067, SRMR = .042, and was a marked improvement in fit compared with both the autoregressive and discrete-time models fit indexes noted previously. To interpret this model, we compared and combined the findings from the direct (cross-lagged) and indirect (mediated) pathways to demonstrate that this model provided the best interpretation of the relations between the parent and child behaviors over time. Indeed, interpreting the cross-lagged paths in the discrete-time model led to the opposite conclusions than were derived by interpreting the total (direct and indirect) effects in the same model. In this model, the direct effects are the coefficients from the cross-lagged pathways in Figure 11.4; the indirect effects are calculated by Mplus (Muthén & Muthén, 2007), taking into account the nonrecursive nature of the reciprocal within-time pathways (Bollen, 1987). The cross-lagged effect path from parent learning support$_{12}$

to child reading $ability_{23}$ is negative, $\beta = -.19$, $p < .0001$ (see Figure 11.4), whereas the total indirect association between these two variables, as mediated through parent learning $support_{23}$ and child reading $ability_{12}$, is positive, $\beta = .48$, $p < .0001$ (see Table 11.1). How did we resolve this apparent discrepancy? We did so by consulting the total effect (which adds together the direct and indirect effects) of the influence of parent learning $support_{12}$ on child reading $ability_{23}$. This total effect is positive, $\beta = .29$, $p < .0001$, which leads to the conclusion that when all three possible pathways from parent learning $support_{12}$ to child reading $ability_{23}$ are considered simultaneously, parents' increases in learning support are found to promote increases in their children's reading ability 2 years later.

The estimates of the direct and indirect effects from child reading $ability_{12}$ to parent learning $support_{23}$ produce similar, seemingly contradictory, findings when they are examined separately. The direct cross-lagged effect is positive, $\beta = .50$, $p < .0001$ (see Figure 11.4), whereas the total indirect effect is negative, $\beta = -.79$, $p < .0001$ (see Table 11.1). However, the combination of the direct and indirect effects produces a significant negative total effect, $\beta = -.29$, $p < .0001$ (see Table 11.1), which supports the conclusion that as children's reading abilities improve from kindergarten to first grade (child

TABLE 11.1
Summary of Total and Indirect Effects from Continuous-Time Cross-Lagged Model

Effects and pathways	*B*	*SE*	β	*Z*
Δ Parent learning $support_{12}$ → Δ Child reading $ability_{23}$				
Total effect	1.10	.05	.29	20.83****
Total indirect	1.82	.12	.48	14.72****
Specific indirect				
1. Δ Parent learning $support_{12}$ → Δ Parent learning $support_{23}$ → Δ Child reading $ability_{23}$	0.90	.09	.24	10.44****
2. Δ Parent learning $support_{12}$ → Δ Child reading $ability_{12}$ → Δ Child reading $ability_{23}$	1.35	.14	.35	10.03****
Δ Child reading $ability_{12}$ → Δ Parent learning $support_{23}$				
Total effect	−.11	.01	−.29	−31.32****
Total indirect	−0.29	.02	−.79	−13.06****
Specific indirect				
1. Δ Child reading $ability_{12}$ → Δ Parent learning $support_{12}$ → Δ Parent learning $support_{23}$	−0.13	.01	−.34	−10.08****
2. Δ Child reading $ability_{12}$ → Δ Child reading $ability_{23}$ → Δ Parent learning $support_{23}$	−0.21	.02	−.57	−11.09****

****$p < .0001$.

reading $ability_{12}$), their parents engage less in learning support in the subsequent 2 years from first grade to third grade (parent learning $support_{23}$).

Together, these findings reveal that both parent effects and child effects are evident in the associations of these constructs over time. As parents increase their provision of learning support to their children, children's reading abilities increase—a clear parent effect. As children improve in their reading over time, parents decrease the amount of learning support they provide—a clear child effect. Both effects are strong and significant and thus are evidence of reciprocal as well as transactional processes. But how can these apparently opposite processes be occurring simultaneously? We interpret these findings as evidence of a scaffolding effect, such that parents' learning support promotes the development of children's reading skills but as children become more proficient in their reading, parents scale down their learning support. In other words, children benefit from the learning support provided by their parents, but they need less of it as their reading skills improve over time.

These nuanced findings clearly could not have been uncovered in a purely discrete-time cross-lagged model. The cross-lagged paths shown in Figure 11.3 are consistent with the direction (although not the magnitude) of the total effect of parent learning support on children's reading abilities (and vice versa) obtained in the continuous-time model. Yet this model leads to the mistaken conclusion that parents' learning support at one time point has a direct, although delayed, effect on their children's reading at a later time point. The decomposition of indirect effects in Table 11.1 reveals very different processes at work, indicating that the effect of increases in parent learning support on increases in children's reading ability is best understood when both direct and indirect pathways are considered simultaneously. It is also worth noting that the cross-lagged paths in the discrete-time cross-lagged model are relatively small (all less than $|.10|$), whereas the total effects in the continuous-time model are 3 times as large, with $\beta s = .29$ and $-.29$ (see Table 11.1). These larger effect sizes further support our contention that the CLPA model represents the relations between parents and children in a more robust and realistic way.

The indirect effect of parent learning $support_{12}$ on child reading $ability_{23}$ occurs through two separate processes. In the first process, increases in parents' learning support across children's kindergarten and first-grade years predict increases in their learning support from $time_2$ to $time_3$ (the fixed-effect interpretation of the autoregressive effect), and this latter increase in turn predicts the increase in children's reading abilities across the same time period (child reading $ability_{23}$). In the second process, increases in parents' learning support across children's kindergarten and first-grade years predict increases in children's reading abilities across the same time period. Early development in children's reading abilities then begets continued growth in their reading abilities from the end of first grade to the end of third grade. These two indirect processes are significant, $\beta s = .24$ and $.35$, $p < .0001$ (see Table 11.1), and both are needed to explain the ways in which parents' learning support predicts children's reading abilities over time.

Two indirect processes are also at work to account for the prediction of change in parent learning support from change in child reading ability over time. Growth in children's reading abilities over the course of kindergarten and first grade predicted a decrease in their parents' learning support over the course of first and second grades, $\beta = -.62$, $p < .0001$, which in turn predicted change in parent's learning support over the second and third grades, as noted previously. Similarly, children's increasing reading abilities in kindergarten and first grade (child reading ability$_{12}$) predicted continued increases in their reading abilities across second and third grades (child reading ability$_{23}$), which in turn predicted less learning support from parents over second and third grades, reflected in an indirect effect of $\beta = -.57$, $p < .0001$ (see Table 11.1). Both estimates of the total effects in Table 11.1 were significant. In each case, the direct and indirect pathways were of opposite sign. This is a common occurrence in CLPA (Oud, 2002) and provides further evidence of the need to move beyond solely discrete-time models to continuous-time models that include both cross-lagged and reciprocal within-time paths.

Parent learning support and children's reading ability are truly transactional processes. The more parents increase their learning support activities over the course of their children's kindergarten and first-grade years, the greater gains their children will have in their reading abilities between the end of first grade and the end of third grade. We also conclude that the more children's reading improved over the course of kindergarten and first grade, the less their parents engage in learning support behaviors in the subsequent 2 years.

The analyses presented here have focused on the parenting context in which children learn to read, yet there are multiple contexts that could be considered using continuous CLPA, such as family socioeconomic contexts. It is also possible to extend the model to include more than two constructs measured repeatedly over time. For example, income has been implicated in both parent investment and in child achievement. We plan to test this possibility in future analyses.

CONCLUSION

In this chapter, we have suggested two new ways in which researchers can estimate and interpret pathways in cross-lagged models that may add richness to their understanding of the transactional relations between parents and children. We have been particularly concerned with developing a better understanding of the short- and long-term processes by which parents and children affect each other, moving beyond a focus on significant cross-lagged effects that can tell only part of the story. Our first recommendation, namely, to include within-time reciprocal paths and to examine indirect pathways through which a parent may affect a child and vice versa, has focused attention on the ways in which such intraindividual growth has predicted growth in the other partner. Our second recommendation, to interpret autoregressive pathways in a fixed effects framework, has enabled us to consider different parameters as indexes of intraindividual stability and change. We hope that our illustrations in this chapter encourage other researchers to test nonrecursive models, particularly when they provide a better match to theories of how individuals develop and mutually influence each other over time. Combined with a fixed effects approach and an emphasis on indirect effects, the continuous-time cross-lagged model represents an important advance in detecting transactional processes in development and human relationships.

REFERENCES

Arbuckle, J. L., & Wothke, W. (1995). *Amos 4.0 user's guide*. Chicago: SmallWaters.

Bandura, A. (1978). The self system in reciprocal determination. *American Psychologist, 33*, 344–358.

Baron, R. M., & Kenny, D. A. (1986). The moderator–mediator variable distinction in social psychological research: Conceptual, strategic, and statistical considerations. *Journal of Personality and Social Psychology, 51*, 1173–1182.

Bell, R. Q. (1968). A reinterpretation of the direction of effects in studies of socialization. *Psychological Review, 75*, 81–95.

Bollen, K. A. (1987). Total, direct, and indirect effects in structural equation modeling. *Sociological Methods, 17*, 37–69.

Bronfenbrenner, U. (1986). Ecology of the family as a context for human development: Research perspectives. *Developmental Psychology, 22*, 723–742.

Burt, S. A., McGue, M., & Iacono, W. G. (2006). Differential parent–child relationships and adolescent externalizing symptoms: Cross-lagged analyses within a monozygotic twin differences design. *Developmental Psychology, 42*, 1289–1298.

Caldwell, B. M., & Bradley, R. H. (1984). *Home observation for measurement of the environment*. Little Rock: University of Arkansas Press.

Cole, D. A., & Maxwell, S. E. (2003). Testing meditational models with longitudinal data: Questions and tips in the use of structural equation modeling. *Journal of Abnormal Psychology, 112*, 558–577.

Davis, J. A. (1985). *The logic of causal order*. Newbury Park, CA: Sage.

Delsing, M. J. M. H., Oud, J. H. L., & De Bruyn, E. E. J. (2005). Assessment of bidirectional influences between family relationships and adolescent problem behavior: Discrete vs. continuous time analysis. *European Journal of Psychological Assessment, 21*, 226–231.

Duncan, G. J., Magnuson, K. A., & Ludwig, J. (2004). The endogeneity problem in developmental studies. *Research in Human Development, 1*, 59–80.

Ferrer, E., & McArdle, J. J. (2003). Alternative structural models for multivariate longitudinal data analysis. *Structural Equation Modeling, 10*, 493–524.

Foster, E. M. (2002). How economists think about family resources and child development. *Child Development, 73*, 1094–1914.

Gershoff, E. T. (2002). Corporal punishment by parents and associated child behaviors and experiences: A meta-analytic and theoretical review. *Psychological Bulletin, 128*, 539–579.

Gershoff, E. T., Aber, J. L., Raver, C. C., & Lennon, M. C. (2007). Income is not enough: Incorporating material hardship into models of income associations with parent mediators and child outcomes. *Child Development, 78*, 70–95.

Gollob, H. F., & Reichardt, C. S. (1987). Taking account of time lags in causal models. *Child Development, 58*, 80–92.

Guo, G., & Harris, K. M. (2000). The mechanisms mediating the effects of poverty on children's intellectual development. *Demography, 37*, 431–447.

Halonen, A., Aunola, K., & Ahonen, T. (2006). The role of learning to read in the development of problem behaviour: A cross-lagged longitudinal study. *British Journal of Educational Psychology, 76*, 517–534.

Haveman, R., & Wolfe, B. (1994). *Succeeding generations: On the effects of investments in children*. New York: Russell Sage Foundation.

Hu, L., & Bentler, P. M. (1999). Cutoff criteria for fit indexes in covariance structure analysis: Conventional criteria versus new alternatives. *Structural Equation Modeling, 6*, 1–55.

Kandel, D. B., & Wu, P. (1995). Disentangling mother–child effects in the development of antisocial behavior. In J. McCord (Ed.), *Coercion and punishment in long-term perspectives* (pp. 106–123). New York: Cambridge University Press.

Kenny, D. A. (1975). Cross-lagged panel correlation: A test for spuriousness. *Psychological Bulletin, 82*, 887–903.

Kessler, R. C., & Greenberg, D. F. (1981). *Linear panel analysis: Models of quantitative change*. New York: Academic Press.

Kline, R. B. (2004). *Principles and practice of structural equation modeling* (2nd ed.). New York: Guilford Press.

Lazarsfeld, P. F. (1940). "Panel" studies. *Public Opinion Quarterly, 4*, 122–128.

Lazarsfeld, P. F., & Fiske, M. (1938). The panel as a new tool for measuring opinion. *Public Opinion Quarterly, 2*, 596–612.

Lerner, R. M. (1991). Changing organism-context relations as the basic processes of development: A developmental contextual perspective. *Developmental Psychology, 27*, 27–32.

Linver, M. R., Brooks-Gunn, J., & Kohen, D. E. (2002). Family processes as pathways from income to young children's development. *Developmental Psychology, 38*, 719–734.

Maccoby, E. E., & Martin, J. A. (1983). Socialization in the context of the family: Parent–child interaction. In E. M. Hetherington (Vol. Ed.) & P. H. Mussen

(Series Ed.), *Handbook of child psychology: Vol. 4. Socialization, personality, and social development* (pp. 1–101). New York: Wiley.

MacKinnon, D. P. & Dwyer, J. H. (1993). Estimating mediated effects in prevention studies. *Evaluation Review, 17*, 144–158.

Martens, M. P., & Haase, R. F. (2006). Advanced applications of structural equation modeling in counseling psychology research. *The Counseling Psychologist, 34*, 878–911.

Muthén, B., & Muthén, L. K. (2007). Mplus (Version 4.21). Los Angeles: StatModel.

Oud, J. H. L. (2002). Continuous time modeling of the cross-lagged panel design. *Kwantitatieve Methoden, 23*, 1–26.

Patterson, G. R. (1982). *Coercive family process*. Eugene, OR: Castalia.

Piaget, J. (1954). *The construction of reality in the child*. New York: Basic Books.

Rathbun, A., & West, J. (2004). *From kindergarten through third grade: Children's beginning school experiences* (NCES 2004-007). Washington, DC: U.S. Government Printing Office.

Rogoff, B. (1998). Cognition as a collaborative process. In W. Damon (Series Ed.) & D. Kuhn & R. S. Siegler (Vol. Eds.), *Handbook of child psychology: Vol. 2. Cognition, perception, and language* (5th ed., pp. 679–744). New York: Wiley.

Rueter, M. A., & Conger, R. D. (1998). Reciprocal influences between parenting and adolescent problem-solving behavior. *Developmental Psychology, 34*, 1470–1482.

Sameroff, A. J., & Chandler, M. J. (1975). Reproductive risk and the continuum of caretaking casualty. In F. D. Horowitz, M. Hetherington, S. Scarr-Salapatek, & G. Siegal (Eds.), *Review of child development research* (Vol. 4, pp. 187–244). Chicago: The University of Chicago Press.

Sameroff, A. J., & MacKenzie, M. J. (2003). Research strategies for capturing transactional models of development: The limits of the possible. *Development and Psychopathology, 15*, 613–640.

Scarr, S., & McCartney, K. (1983). How people make their own environments: A theory of genotype → environment effects. *Child Development, 54*, 424–435.

Schafer, J. L., & Graham, J. W. (2002). Missing data: Our view of the state of the art. *Psychological Methods, 7*, 147–177.

Sigel, I. E., & Parke, R. D. (1987). Structural analysis of parent–child research models. *Journal of Applied Developmental Psychology, 8*, 123–137.

IV

NEW DIRECTIONS

12

TRANSACTIONS AND STATISTICAL MODELING: DEVELOPMENTAL THEORY WAGGING THE STATISTICAL TAIL

RICHARD GONZALEZ

What new statistical techniques are available for developmental psychology? What hot new technique is available for me to use in my research? Such questions may appear reasonable, but I believe they place emphasis on the wrong attributes. I prefer to reframe such questions. What new theoretical and empirical properties are developmental psychologists testing and measuring? This changes the focus from asking about new statistical techniques to asking about theory, modeling and testing. It highlights the need to challenge current methodological assumptions and methods rather than to find ways to tweak existing methods to fit new empirical and theoretical questions.

In this chapter I focus on the key features of the transactional model. Some of these features require the development of new methodological techniques that provide appropriate tests. Once the key features of the transactional model are reviewed, I outline several statistical developments that facilitate the implementation of these features in statistical models. My hope is that this chapter will inspire developmental psychologists to think more broadly about the kinds of questions they ask and will inspire methodologists to think more broadly about the statistical techniques they develop.

I thank Arnold Sameroff for his illuminating discussions over several lunches about the transactional model.

BACKGROUND

When thinking about the use of a mathematical or statistical model, one might find it useful to make an analogy to a road map. A road map is helpful to the motorist driving between two cities. The map informs the motorist which streets and highways to take, which is typically the key information the motorist wants to know. A road map does not include every detail, nor is it a perfect replica of the original. The map is a simplified representation that contains only the features necessary for its intended use. The typical road map by itself would not be useful, for instance, in the study of social networks in a city. The road map was not designed to offer information about social networks, though one can imagine different kinds of maps that could be constructed by a researcher studying social networks. A modified road map that overlays key individuals and their social ties could be useful when the goal is to understand the location of the critical nodes of a social network in a geographical layout.

A statistical or mathematical model is analogous to a road map because it is a simplified, abstracted representation of a more complex phenomenon. The mathematical model is a useful representation in so far as the abstraction contains the features that are necessary to understand the phenomenon. Take the simple general linear model $Y = X\beta + \varepsilon$. This model is a simplification in the sense that a dependent variable Y is modeled as a linear combination of predictor variables X (where X is a matrix, including the unit vector that models the intercept) plus a stochastic term ε. One learns in introductory statistics courses (a) to check the assumptions of this model, (b) how residuals can diagnose problems with the equality of variance and normality assumptions, (c) that measures of influence can be used to check for outliers, and (d) to deal with curved data by transforming the dependent variable or the predictor(s) or by including polynomial terms as additional predictors. The simple general linear model can be extended further into other types of statistical distributions such as the generalized linear model that encompasses logistic, probit, Poisson, gamma, and other forms of regression. One learns (a) to extend the simple linear model into one with random-effects terms and multilevel equations that permit modeling different trajectories for each individual and (b) that the simple linear model can be extended to nonlinear regression and that it is even possible to mix and match elements from these various generalizations, such as a multilevel nonlinear regression equation with a Poisson distribution.

There are trade-offs in any modeling exercise. A model can be made more complex, and hence more representative, but it comes at the price of being harder to understand and less parsimonious. As a model is endowed with more complexity, the better it can fit a particular data set. Although complexity makes a model richer, it also makes it more likely that idiosyncratic features of the data dominate the fit, thus reducing the generalizability of the statistical conclusions that emerge from the estimation exercise. Navigating such trade-offs requires trial and error and a solid appreciation of the intended goals of the modeling exercise (i.e., the intended use of the road map).

Some people claim that a mathematical model forces one to be precise. I agree that precision is an important feature of a mathematical model. A lesser known property of mathematical models is that they allow one to be more general in the following sense. If a mathematical property is shown not to represent a phenomenon adequately, one also rejects every theory (not just current theory but theories yet to be developed) that makes use of the mathematical property that has been rejected by data. The usual empirical approach is to test theories one at a time. As new theories emerge, new empirical tests are needed. Today one tests a competitor theory against the old standard theory. If tomorrow a new theory emerges, one needs a new empirical test for the new contender. However, by focusing on properties of theories and testing those properties, one may make more general empirical and theoretical statements that cut across an individual study or test.

A good example of such generality from mathematical modeling occurs in research on similarity judgment. For a long time, the literature focused on geometric psychological representations that could account for similarity judgments. If a child judges two items as similar, those items should be represented as close in "psychological space." Two items judged as dissimilar could be represented as relatively far in psychological space. Through a series of judgments, researchers hoped to characterize the nature of psychological space. Such representations could be used, for example, to study the developmental trajectory of knowledge and meaning. Researchers struggled to find a mathematical representation to account for similarity judgments. Tversky (1977) showed that similarity judgments do not always obey key properties that characterize all distance metrics—symmetry and triangle inequality. *Symmetric judgments* occur when a child judges the similarity of Object A to B to be the same as the similarity of Object B to A (order of judgment is irrelevant). *Triangle inequality* can be viewed as a type of geometrical constraint: If a third Object C is "between" Objects A and B, then the direct distance between A and B cannot exceed the sum of the intermediate distances AC and CB (e.g., the distance between Seattle and Boston cannot be greater than the sum of the distances of Seattle to Ann Arbor and Ann Arbor to Boston). If those two properties are violated, the debate about which distance metric to use is silly because a distance metric for such data cannot exist. It would not be an issue of how well the model fits the data, but the model could be rejected outright for not satisfying the necessary properties of symmetry and triangle inequality. Not just existing theories but any theory yet to be developed that makes use of symmetry or the triangle inequality would be in jeopardy if either of those properties were found not to hold. It is not necessary to conduct new empirical studies to test new theories that make use of symmetry and triangle inequality. Once the empirical boundary conditions of those conditions are known, then one also knows something about the boundary conditions of any theory that makes use of those properties. Tversky (1977; also Tversky & Gati, 1982) provided convincing evidence of violations of both symmetry and the triangle inequality. This evidence provides one example of the benefits of a clear mathematical or statistical model. A good mathematical model can be simultaneously more precise and more general than the standard empirical approach.

There is much benefit from a dual approach that focuses on clear theoretical statements of psychological phenomena and the implementation of those statements in a mathematical or statistical model. In this chapter I highlight some key properties of the transactional model and discuss possible implementations of those properties in formal models. My discussion remains at the general level of process rather than at specific hypotheses about particular variables.

TRANSACTIONAL MODEL

Given my emphasis on having a clear theoretical statement before pursing statistical details, I need to outline the transactional model. There are several published pieces outlining, expanding, and applying the transactional model, so I refer the reader to various sources (Sameroff, 1995, 2000; Sameroff & Chandler, 1975; Sameroff & MacKenzie, 2003). The other chapters in this volume also provide excellent descriptions, illustrations, applications, and extensions of the transactional model. I outline the key pieces of the model that I believe are critical to specify a data analytic framework.

One definition of the transactional model comes from Sameroff (1995): "In the transactional model the development of the child is seen as a product of a continuous dynamic interaction of the child and the experience provided by his or her family and the social context" (p. 663). Sameroff and MacKenzie (2003) stated that

> transactions are documented where the activity of one element changes the usual activity of another, either quantitatively, by increasing or decreasing the level of the usual response, or qualitatively, by eliciting or initiating a new response, for example, when a smile is reciprocated by a frown. (p. 617)

The transactional model has several key properties, including multiple observations (time) over multiple variables of several interacting individuals. The nature of the social interaction creates a context that can also influence the variables. In this sense some variables are *endogenous* because they are response variables, which could also serve as predictor variables. The transactional model allows for heterogeneity of response (people differ from each other); it takes a dynamic perspective because it focuses on change over time; it is path dependent and nonlinear. It posits bidirectional reciprocal relationships between variables. The multivariate structure of the transactional model is rich. It includes person variables, genetic variables, biological variables, societal variables, environmental variables, parental and caregiver/socializer/teacher variables, cultural variables, neighborhood variables, and economic variables, as many of the chapters in this volume illustrate.

In this chapter I outline a few statistical properties that are relevant to implementing the transactional approach in an analytic framework, although I cannot review every statistical detail related to the transactional model in this short chapter. I extend here the article by Sameroff and MacKenzie (2003) that couched the transactional model within a few standard statistical models. Throughout the chapter I assume normally distributed interval data though at some points mention areas in which both models for other distributions (such as Poisson for counts) and extensions to nonparametric models exist. The chapter is organized around the key properties of the transactional model—a multivariate, dynamic, endogenous, heterogeneous, path-dependent system involving multiple individuals.

DESCRIBING MULTIVARIATE PROCESSES OVER TIME

One needs to be able to visualize the phenomenon under investigation. Psychologists typically do not make as much use of graphs as they should (a strong case for the use of graphs in regression models was made by Gelman & Hill, 2007). They are familiar with graphs to diagnose models (such as using residual plots in regression analyses to detect outliers, to check the equality of variance assumption, and to check the normality assumption). Psychologists also use graphs to plot means and trends over time and to highlight details in multivariate time series.

In multivariate time series with k variables in t time points, one observes a data matrix for a single subject structured as

$$\begin{bmatrix} y_{11} & y_{12} & \cdots & y_{1t} \\ y_{21} & y_{22} & \cdots & y_{2t} \\ \vdots & \vdots & \vdots & \vdots \\ y_{k1} & y_{k2} & \cdots & y_{kt} \end{bmatrix}, \qquad (12.1)$$

where each row denotes a different measured variable and each column a different time. In general, there may be missing cells in this matrix (e.g., the participant decided not to respond to a variable at a particular time, the researcher chose to omit the variable at a particular time, or some event occurred prohibiting an observation).

The standard plot most people construct has time on the horizontal axis and the dependent variable on the vertical axis, thus displaying the dependent variable as a function of time. Curves can be plotted for individual subjects to illustrate trajectories. Psychologists tend to take each row of the data matrix (a single variable for a single subject over *t* time points) and plot each row against time. Recent advances in latent growth curve analyses wrap statistical theory around the fundamental properties of such a plot (e.g., random effects on intercept and slope terms to handle heterogeneity).

A different type of plot shows multiple variables changing in time on the same plot. That is, multiple rows in the previously described matrix are plotted on the same graph. One way to accomplish this is by plotting variables against variables (as opposed to variables against time) with time as an implicit variable. For instance, Figure 12.1 shows a plot for an individual subject's meals throughout a particular day. The complete variation of the percentage of the meal's carbohydrate, fat, and protein content (which sum to 100%) creates a triangle on which each vertex corresponds to a meal with 100% of that component. For instance, the point (0%, 100%) corresponds to a meal (unrealistic as it may be) that is 100% fat. The other two vertices denote meals that consist of 100% of the other two components. A point in the interior of the triangle corresponds to a meal that has some combination of all three components. To use a concrete example, a meal that is ⅓, ⅓, and ⅓ for each of the components is depicted as the point (33%, 33%) in the triangle. The three meal components sum to 100%, so only two axes are needed to display all three components (the third component is implicit in the graph). I chose to make carbohydrate the implicit component in the graph.

The plot can display more information. For instance, the size of the point can be related to the number of calories, so together the location of the point in the triangle and the size of the point inform the viewer about the key elements of the meal (composition and calories). Time can be depicted as the trajectory of points within the triangle. In this way the meals (including time) can be represented in the triangle. On a computer screen, the graph can be animated so that the trajectory over multiple days can be visualized relatively easily. My collaborators Grazyna Wieczorkowska and Malgorzata Siarkiewicz and I have used this representation to understand actual diets of research participants over several weeks. We have also used color to code points to reflect a sixth variable such as self-reported hunger (the other five variables are percentage of carbohydrate, protein, and fat of the diet; the number of calories; and time). By studying how the points move around the triangle and lagging the hunger variable various hours prior to and after the meal, we can examine, for instance, how hungry a participant feels after a predominantly high-protein meal with low calories compared with a high-calorie predominantly

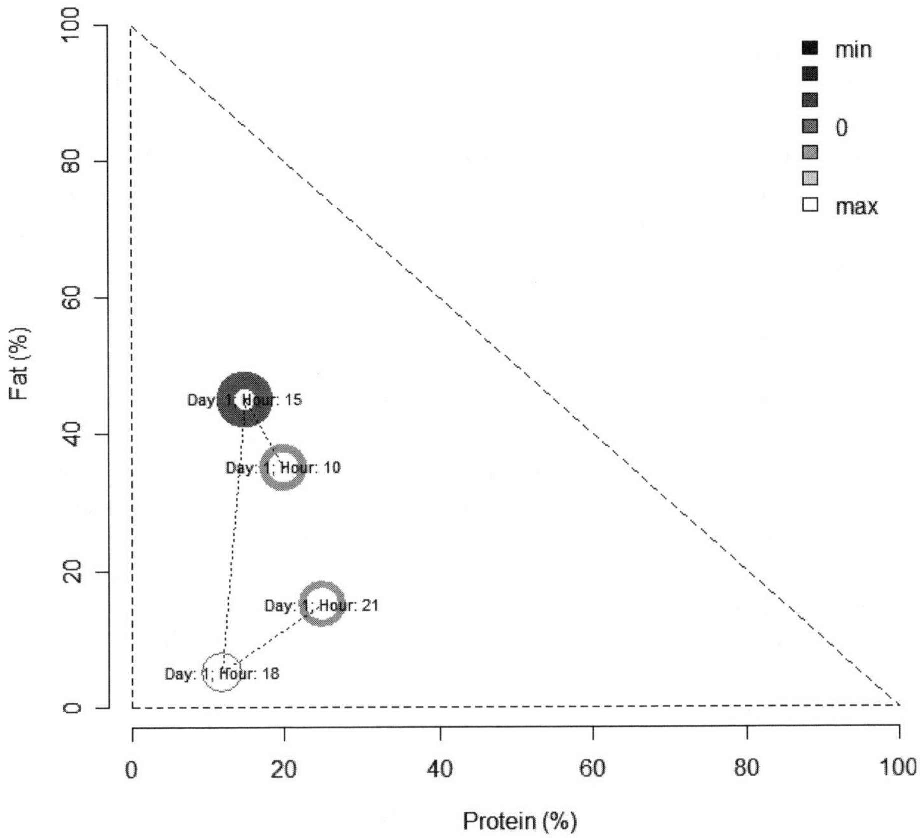

Figure 12.1. Triangle plot depicting four time-stamped meals in one day; the percentage of fat, carbohydrate, and protein for each meal; the number of calories (related to the size of the circle); and the hunger (related to the gray scale of the point) for one subject.

high-carbohydrate meal. Such relationships between multiple variables over time would be relatively difficult to detect using the standard graph with time on the horizontal axis and different curves for each variable. Of course, there remains the problem of adding a statistical conclusion such as a hypothesis test to this graph, but the value of visualizing a dynamic process over multiple variables cannot be underestimated for developing intuition about the processes that underlie the phenomenon. As one adopts more complicated statistical models that allow for individual differences (i.e., different trajectories for each subject), it becomes important to visualize such data and represent the patterns descriptively to achieve a deeper understanding of the underlying processes. A well-designed plot can complement a table of variance component estimates from a multilevel statistics package, making it much easier to understand the finding.

The diet example involves a plot with time as an implicit attribute that is depicted by the movement of the points. A related plot has been used in the study of social engagement between parent and child (Hollenstein, Granic, Stoolmiller, & Snyder, 2004). There are different types of plots that are also useful with multivariate dynamic data, such as plots that illustrate the relation between a variable and its first derivative (e.g., Boker & McArdle, 1995) and plots of parameters, which can facilitate model comparison (e.g., Pitt, Kim, Navarro, & Myung, 2006).

Statistical and Software Implementations

The computer package R has several plotting commands for visualizing trajectories. The lattice package in R produces elegant graphs using modern principles of graph design. The diet plot in Figure 12.1 was generated with an R function that I wrote. A similar triangle plot can be found in the R package plotrix. MlWin, a commercial package for multilevel modeling, has capabilities for plotting trajectories in the context of multilevel models. Other types of useful graphs are vector field graphs that can be computed in packages such as Maple and Mathematica or specialized programs (e.g., Boker & McArdle, 1995).

Researchers sometimes counter with, "Looking at graphs is a great idea, but how do I convey them in a paper?" Obviously, it is not possible to include every graph in a paper, and animated graphs cannot be depicted on the printed page. Key findings, though, need a corresponding table or graph. Consider how much more informative an analysis of variance (ANOVA) can be when accompanied by a table or graph of means. As researchers move into fancier statistical techniques, including growth curve analysis, multilevel models, and dynamic processes, it becomes necessary to develop corresponding methods of displaying descriptive information. Omitting those descriptive features from such fancy analyses detracts from the overall informativeness of the analysis in much the same way as omitting the table of means detracts from the informativeness of an ANOVA source table.

HETEROGENEITY

People differ. Many psychologists give lip service to individual differences; a few psychologists spend a major part of their careers understanding how people differ. Heterogeneity is one of the observations that led to the development of the transactional model. Why is it that not all kids with brain damage develop problems later in childhood? Differences in people are now seen as important and as providing clues about key elements of the underlying process. Psychologists need to use statistical analyses that (a) make it straightforward to estimate and describe how people differ; (b) do not make unrealistic assumptions that all people have the same degree of responsiveness to a manipulation; (c) make it easy to find the predictors of such individual differences; and, if needed, (d) allow variability to be a predictor of other variables.

One freeing aspect of modern statistics is that it is no longer necessary to make simplifying assumptions that all participants within a treatment group are the same or respond to treatment the same way. It is relatively easy to extend the standard statistical models so that heterogeneity is permitted in the parameters—a statistical way of saying that there are individual differences. Coupled with a multilevel model approach, or a Bayesian approach to statistical estimation and inference, it is fairly routine to proceed with statistical models that encompass heterogeneity.

The idea underlying a latent growth curve is that there is a common curve that characterizes a general trend, but there is variability across individuals. Each person can be endowed with his or her own curve, modeled in an efficient way, on which interindividual differences are examined simultaneously with intraindividual differences. This follows from new developments in random-effects analysis, multilevel modeling, and also structural equation modeling. The curve is decomposed into key components such as a slope and intercept in the case of a straight line trajectory, and those components are treated as random effects that permit the modeling of heterogeneity. To me, this is the major benefit that random-effects models offer—they allow one to model heterogeneity. As a side benefit, they also allow a more solid generalization to the population (which is the aspect that many in the field believe is the key benefit of random-effects models).

Latent class models allow one to find groups of individuals who respond similarly. Intuitively, if we take points in parameter space, then individuals with similar parameters form a cluster. Latent class analysis and its extensions deal with the problem of finding such clusters of parameters, such as which subjects have similar slopes and intercepts.

Rather than finding clusters, latent growth curve analysis allows each subject to have his or her set of parameters. Intuitively, one can view latent growth curve analysis as a latent class with N clusters (i.e., one cluster for each subject).

Sameroff and MacKenzie (2003) pointed out that although there may be variability in the individual observations and traits, there may be *constants* in the types of processes that maintain the relation between the individuals and the context. One of the obvious constants is that the person is part of the context. I am impressed with the politeness of many undergraduates who hold the door open or hold the elevator for me as I approach. This could be a general trait of undergraduates or it could be that the graying, middle-aged man who looks like a professor is a common element in those social interactions between me and the undergraduates.

Statistical and Software Implementations

Perhaps the simplest model that includes a form of heterogeneity is the simple paired *t* test. Indeed, every repeated measures ANOVA includes heterogeneity. I illustrate with the simple paired *t* test. The standard parameterization for this model is

$$Y_{ij} = \mu + \alpha_j + \pi_i + \varepsilon_{ij}, \tag{12.2}$$

where each observation Y_{ij} for subject i at time j is modeled as a function of four terms: a constant μ for all subjects at all times, a fixed-effects time parameter α that codes the main effect for the difference between Time 1 and Time 2, a random-effects subject parameter π that codes the main effect for subject, and the usual error term ε. The π terms model individual differences and form the key aspect of what people talk about when they say that a repeated-measures ANOVA can be more powerful than a between-subjects ANOVA because it controls for individual differences. Unfortunately, the standard application of a repeated-measures ANOVA treats individual differences as something to control to reduce the error term. This is not a limitation of the statistical technique but a constraint imposed by how the model is typically used and how it is interpreted. It is possible, as one sees in multilevel models, to embrace individual differences as something to understand, model, and predict.

Latent class analysis can be implemented in various statistical programs. Nagin (1999) has SAS macros for latent class analysis of growth curves. The package flexmix in R allows for the fitting of latent classes in general regression contexts. Latent growth curves can be implemented in standard multilevel programs (e.g., HLM, MlWin) as well as structural equation programs (e.g., LISREL, EQS, AMOS). The statistical package Mplus provides much flexibility for incorporating latent class in standard multilevel and structural equation models. Gibbons and Hedeker (see http://tigger.uic.edu/~hedeker/mix.html) have developed specialized programs for random-effects models for advanced regression such as survival analysis and ordinal regression.

The Bayesian approach provides a natural way to implement multilevel models. Once the analyst assumes a distribution over parameters, then the analyst is in the domain of multiple levels. Bayesian implementations, however, still require some programming on the part of the user for all but the simplest analysis problems. A good Bayesian software program is WinBUGS, and there are several packages in R that implement Bayesian approaches (e.g., MCMCpack). The package MLWin provides a simple Bayesian estimation approach to multilevel models.

NONLINEARITY

Trajectories need not be straight lines. There may very well be important psychological information in the nonlinearity. At what age does a process begin? At what age does the process accelerate or produce a dramatic change in another variable? At what age does the curve hit asymptote? Once one can parameterize such questions, one can ask second-order questions such as what variable predicts the point at which acceleration begins and what processes control the time at which the asymptote occurs? In other words, once one operationalizes a psychological property as a parameter and treats it as a random variable to allow for heterogeneity, then one can predict that parameter (as in any multilevel model). The additional concept here over the previous section of this chapter is that I am now dealing with parameters from a nonlinear representation.

Most psychological researchers who model nonlinearities restrict themselves to polynomials. The usual statistical advice is as follows: If something is not a straight line, try a quadratic. If that does not work, try a cubic, and so on, until you get a good fit. This is a mindless way to model curvature because it merely models the number of bends in the curve—no bends is linear, one bend is quadratic, two bends is cubic, and so on, with a *bend* being an inflection point in the curve or a point having a first derivation equal to zero.

A natural extension of polynomials is to model nonlinearity directly through nonlinear functions. There is much work in mathematical psychology in this regard, for example, in estimating models of reaction time distributions, decision making, and performance (e.g., Busey & Loftus, 1994; Gonzalez & Wu, 1999; Rouder, Lu, Speckman, Sun, & Jiang, 2005; Wu & Gonzalez, 1996). The model by Busey and Loftus (1994) is relatively simple to illustrate without getting into the experimental details. The problem is digit recall when strings of four digits are presented for very brief durations. The performance curves (chance corrected) fit quite well with a nonlinear function of the following form:

$$p(t) = \begin{cases} 0 & (t < L) \\ Y\left(1 - e^{-(t-L)/c}\right) & (t \geq L) \end{cases} \tag{12.3}$$

where p is the proportion correct, t is the exposure duration of the stimulus, c is the exponential growth constant, L is the maximum duration that gives chance-level performance, and Y is asymptotic performance (which Busey & Loftus, 1994, set to 1 in their article).

The interesting point about this functional form is that the parameters have psychological meaning. Specifically, L is interpreted as the duration at which the curve begins to move away from chance performance. In other words, the value of L is the time at which the process begins. The parameter c indicates the rate of change; the parameter Y indexes the asymptotic level of performance. It is possible to allow for heterogeneity by incorporating a random-effects approach to the estimation, thus allowing each subject to have his or her own values for the three parameters L, c, and Y. In this way, the curves are modeled with parameters that provide relevant psychological information. Contrast this approach with the more typical polynomial regression approach that merely indicates whether trajectories are linear, quadratic, cubic, or some higher order (though compare with Cudeck & du Toit, 2002, who reparametrized the quadratic into more interpretable and psychologically relevant parameters).

There is a different sense of nonlinearity that is also relevant to the transactional model. This form is related to dynamic systems theory and involves, for example, models that are sensitive to starting configurations and deal with nonlinear differential equations. There has been some attempt to use dynamic systems theory in developmental psychology but a complete review of this approach is beyond the scope of this chapter (see, e.g., Granic & Hollenstein, 2003).

Statistical and Software Implementations

Most statistical packages allow nonlinear estimation techniques, including the popular commercial packages SPSS and SAS. There are plenty of new developments underway in the statistical community, including generalized nonlinear regression that works for data that follow generalized exponential distributions such as binomial, Poisson, gamma, and negative binomial. Within the R program, two relevant packages are lmer and nlme.

There are also nonparametric techniques such as splines that can be used to model nonlinearities. There have been new developments within the Bayesian approach to include a random-effects approach to splines (i.e., each subject is allowed to have a different number of knots as well as different knot locations). An excellent example of this work is found in Kim, Menzefricke, and Feinberg (2007). In this way it is possible to have interpretable parameters (knots) of subject trajectories, allow for nonlinearities through splines, and allow for heterogeneity through random effects in the parameters.

MULTIPLE INDIVIDUALS

People influence each other. Together, people create contexts that are emergent and influence the behavior, thoughts, and feelings of individuals in the group. There are processes related to both interdependence and similarity that can occur with multiple interacting individuals. My collaborator Dale

Griffin and I have written about these different approaches to modeling dyadic and group data (Gonzalez & Griffin, 2001). Interdependence and similarity challenge the usual statistical assumption of independence by introducing correlations across people. Statistical models of social interaction include interdependence and similarity as psychological parameters that are modeled rather than treated as nuisance variables that need to be eliminated or controlled. For a recent book-length treatment, see Kenny, Kashy, and Cook (2006).

For a long time, interindependence of data was treated as a nuisance that needed to be eliminated or corrected. The focus had been on the nasty effect violating independence had on standard errors and tests of significance. The newer models change the emphasis of interindependence from one of nuisance to showcasing the interdependence as a relevant psychological property that needs to be captured, modeled, and understood (Gonzalez & Griffin, 2000).

One notion of interdependence can be captured using the actor–partner model (e.g., Cook & Kenny, 2005). The actor–partner model involves an extension of the standard cross-lagged regression path model to a situation of multiple variables over multiple people. Figure 12.2 depicts a simple model for distinguishable dyad members. The stability coefficients (horizontal paths *a* and *d*) reflect the relation between two variables for the same person, partialing for the cross-path (i.e., in the case of the dyad, for the other person's predictor variable). Likewise, the cross-paths (paths *b* and *c*) represent the influence of one person's predictor on the other's criterion variable, partialing

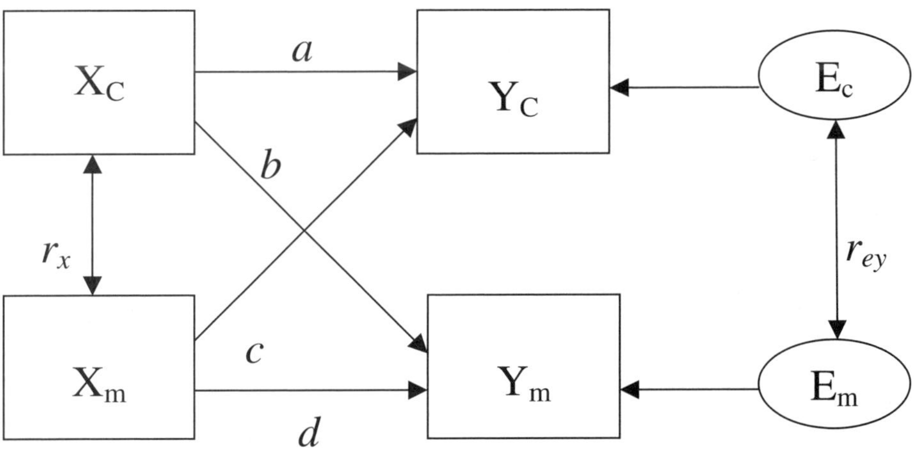

Figure 12.2. Actor–partner model. The focus is on partialing out one's own influence on different variables from the influence the partner has on one's own variables. Horizontal paths *a* and *d* reflect the relation between two variables for the same person, partialing for the cross-path. The cross-paths *b* and *c* represent the influence of one person's predictor on the other's criterion variable, partialing for the stability coefficient.

for the stability coefficient. A simple example is that a predictor variable is measured for both the mother and the child (X_m and X_c). A second variable is also measured for both the mother and the child (Y_m and Y_c). Errors are correlated because individuals are members of the same dyad or group. The model is modified slightly when the two individuals are exchangeable, such as in same-sex couples or same-sex siblings (D. Griffin & Gonzalez, 1995), and can be extended to groups (Gonzalez & Griffin, 2001).

A different type of interdependence is based on similarity. The latent variable model captures similarity, or shared variance, at the dyad level as distinct from unique variance at the individual level. The latent variable model allows for correlations across the dyad-level latent variables (group-level correlation) as well as correlations across the unique variance for the same individuals (individual-level correlation). As shown in Figure 12.3, in the case of two variables for each dyad member, one can model the shared variance on each person as well as each dyad. As with the actor–partner model, modifications to exchangeable cases and groups have been made (Gonzalez & Griffin, 2001; D. Griffin & Gonzalez, 1995).

Statistical and Software Implementations

Models for nonindependence across multiple individuals can be implemented in multilevel models and in structural equation models as well as in simple Pearson correlations once the data are organized in an appropriate manner (e.g., Gonzalez & Griffin, 1999; D. Griffin & Gonzalez, 1995; Olsen

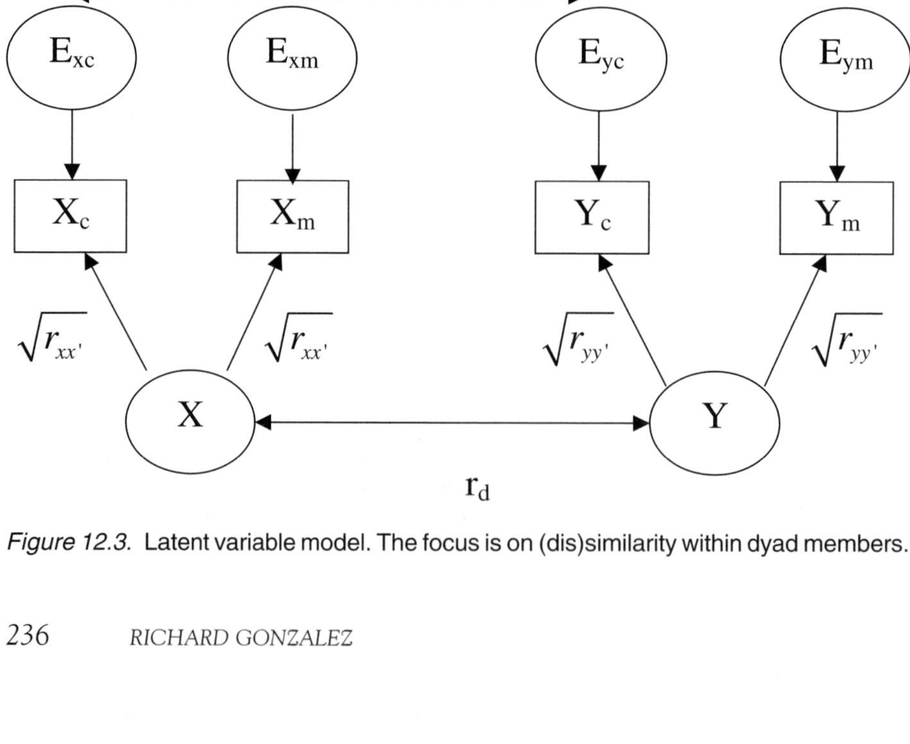

Figure 12.3. Latent variable model. The focus is on (dis)similarity within dyad members.

& Kenny, 2006; Woody & Sadler, 2005). One can take a multilevel approach that individuals are nested within the dyad (or group). Thus, interdependence can be captured and modeled through the nesting procedure. One implementation of this approach assigns a different linear equation to each person and estimates covariances of parameters across equations, that is, the covariance of the same parameter over two people (Barnett, Marshall, Raudenbush, & Brennan, 1993). Or, equivalently, one may take a structural equation modeling approach and add another layer of latent variables to represent the group as shared variance from the individuals who compose the group (e.g., Gonzalez & Griffin, 1999; Kenny & la Voie, 1985). There are additional complications in implementing analyses in statistical packages, such as whether the dyad members are exchangeable or distinguishable. The extension to longitudinal dyadic or group data analysis necessary for testing transactional models is still in its infancy, though for preliminary suggestions see (Gonzalez & Griffin, 2002). W. Griffin (2000) provided a coordination index for multiple individuals along with ideas for visualization and animation of the process. Another interesting extension is the added richness that can emerge from data that have a round-robin structure (e.g., Cook & Kenny, 2004). Much progress is needed though to capture the interdependent processes that are posited in the transactional model—multiple people across multiple variables with bidirectional, reciprocal relations.

ENDOGENEITY

The property of endogeneity may be one of the most important components of the transactional model. Usually, an *endogenous variable* is defined as a variable that is modeled as inside the system being studied. An endogenous variable can be observed or latent; it is predicted by other variables, and it can also predict other variables. By contrast, an exogenous variable is a variable outside the system being studied and so cannot be modeled as a dependent variable within the system.

The transactional model posits that multiple people influence each other across multiple variables over time. Hence, one individual influences another and vice versa. This leads to problems of bidirectionality as discussed by Sameroff and MacKenzie (2003). There are also corollary processes to endogeneity. There are processes that are emergent within the system. *Emergent* properties are those that are not necessarily observable at, say, the individual level but are observable at a higher level of analysis such as the group level. The hydrogen atom has particular characteristics, as does the oxygen atom, but the H_2O molecule has emergent properties that are not easy to predict from only the properties of the individual atoms. Psychological examples of emergent properties are constructs such as norms and groupthink, which can be modeled as group-level processes that are not so easy to understand from merely knowing the properties of individual group members.

Economics has been dealing with endogeneity for a long time. Theoretical relations between supply and demand are endogenous and bidirectional. Demand changes supply; supply changes demand. In economics, such processes are modeled through simultaneous equations. Demand may be influenced by supply along with a set of predictor variables; supply may be influenced by demand along with a set of (not necessarily overlapping) predictor variables. A body of econometrics is devoted to these kinds of bidirectional models (e.g., Haavelmo, 1943; Yang, Chen, & Allenby, 2003).

Endogeneity characterizes the idea that although the system influences me, I also influence the system because I am part of the system. That is, I am part of and define the very system that is influencing me. This is a difficult concept for researchers brought up in the sophomoric tradition of an independent variable influencing a dependent variable. As Sameroff and MacKenzie (2003) pointed out, one needs to understand both the parents' role in the child's behavior as well as the child's role in eliciting the parents' behavior (see also chap. 11, this volume). The physicist Robert Savit and I, in collaborative research, modeled the decision-making behavior of individuals in a group in which the outcomes given to each group member depended on the group decision (Savit, Koelle, Treynor, & Gonzalez, 2004). The outcome is created by the decisions of each group member. Many aspects of social cognition, for example, that borrow ideas blindly from cognitive psychology to model social psychological processes miss the key idea that in social interaction the social stimuli that make up the cognitive machinery "look back" and interact with the social perceiver. Such an interaction with stimuli and subject is not usually possible in the standard cognitive study, and attempts by researchers to apply such models from cognitive psychology to social interaction when the possibilities for endogeneity are high are likely to fail in modeling complicated social interaction (Ickes & Gonzalez, 1994).

Statistical and Software Implementations

Structural equation modeling can handle endogeneity and also the simultaneous equation problem (e.g., supply and demand). One merely sets up a linear equation for demand and a linear equation for supply (with supply as one of the predictors for demand and demand as one of the predictors for supply). Each equation has its own error term, and the error terms are correlated because the two equations are estimated from the same data. Details of the implementation can be found, for instance, in the documentation to PROC CALIS, which is the structural equation modeling procedure built into SAS. In this way bidirectionality can be incorporated into the structural equation framework. There are identification difficulties one may encounter, so bidirectionality is not as straightforward as merely adding two arrows in opposite directions to the usual "square–circle–arrow" structural equation graph. Some of these difficulties are discussed in path analysis and structural equation modeling textbooks such as Kenny (1979).

DYNAMIC PROCESSES

People change. It is obvious that time is a necessary aspect of change. To understand a dynamic process, though, one typically needs more than two time points. As Rogosa, Brandt, and Zimowski (1983) pointed out in their defense of difference scores to measure change, the difference of two time points can be interpreted as a slope. But to have a rich data source to test dynamic processes it is necessary to have more than two time points.

There has been recent interest in nonlinear dynamical systems (e.g., Durlauf & Young, 2001; Gottman, Murray, Swanson, Tyson, & Swanson, 2002; Nowak & Vallacher, 1998). These analytic tools promise to provide much insight into dynamic processes. However, a major chunk of the work has been isolated to computer simulation and has focused on studying how the parameters of the model in different combinations produce different observable behavior. Much knowledge can be gained from this approach if used properly, but it is not the standard data-fitting exercise. Indeed, I am aware of very few cases in which model fitting in the sense of using data to find the best fitting set of parameters in a dynamic system is a tractable problem. In many cases, the parameters themselves are highly intercorrelated, making a general estimation routine intractable.

Statistical and Software Implementations

Time-series analysis is a standard analytic approach that psychologists use with repeated measures. A major focus of this approach is the autocorrelation due to repeated measurement. In some cases this autocorrelation is treated as something that needs to be corrected, whereas in other cases the temporal correlations are the exciting aspects of the data to mine. As mentioned in an earlier section, Boker and McArdle (1995) provided one attempt to use data to analyze graphically the properties of dynamic systems. Developing statistical analyses of dynamic systems that include data fitting and parameter estimation is still, in my opinion, an open area of research. Boker and Nesselroade (2002) provided insights into dynamical systems and discussed the challenges in implementing statistical estimation with few time observations and in the presence of measurement error.

PATH DEPENDENCE

It matters where I came from. The property of endogeneity reviewed in the previous section produces a further nuance: Not only do variables influence each other, but there are processes that constrain future responses and outcomes. If a child throws a temper tantrum, the parent has various ways to respond, but the act of the temper tantrum constrains the parent's potential responses. Likewise, once the parent selects a response to the temper tantrum, that response constrains or affords subsequent behaviors of the child. Such a process is represented in Figure 12.4. The child performs an action *a*. The parent has several responses and chooses one (depicted by the solid line segment). An optimal response for that situation is denoted with an asterisk. Following the solid line path from left to right (time), one sees that the child has several actions at *g*, not all of which may be optimal. If the child's choice depends only on the previous action *f* by the parent, without consideration of how the situation arrived at *f* (e.g., either through *e* or *e**), then the system is said to be path independent (i.e., satisfies the Markov property). However, if the child's choice depends on the particular path in getting to *f*, then the system is said to be path dependent. Heckman's (2005) framework of causality could handle such processes through selection in the sense that parent's

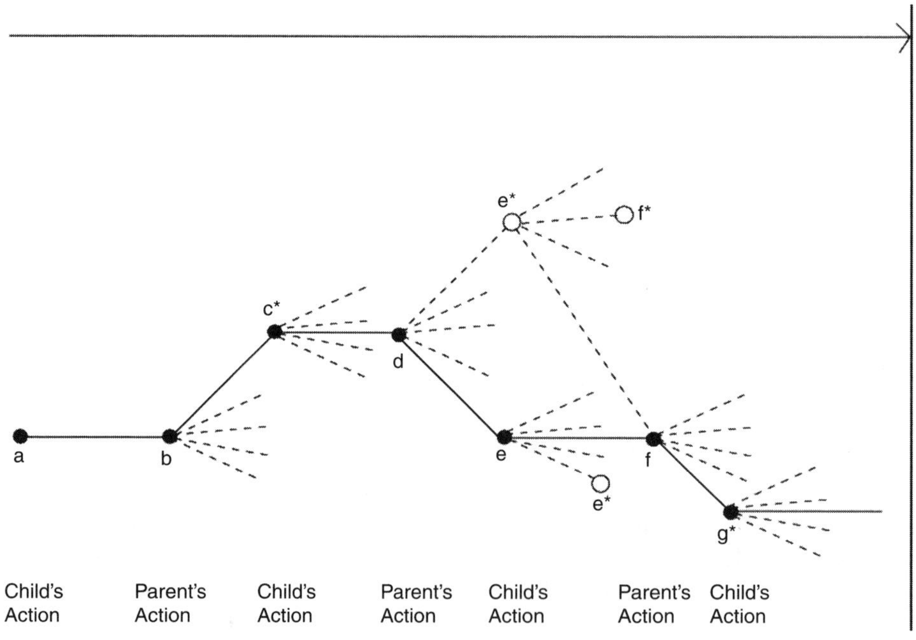

Figure 12.4. Sequence of child–parent behaviors in which each behavior provides constraints for the next turn. An asterisk denotes an optimal response.

choice assigns constraints to child's behavior, child's behavior assigns constraints to parent's behavior, and so on.

Contemporary researchers in computer science and neuroscience have rediscovered concepts from the learning literature (e.g., Hebbian learning) and have developed new models called, broadly, *reinforcement learning models* (Sutton & Barto, 1998). Most of these models assume the Markov property of path independence, but these models show promise for allowing the modeling of path-dependent processes. I am aware of some applications of reinforcement learning models in economic behavior and decision making (Erev & Roth, 1998; Salmon, 2001). There is opportunity to develop statistical models that permit testing of path-dependent transactional processes.

Statistical and Software Implementations

A very promising statistical approach in the spirit of such models that permit path dependence is the state space approach developed by Granic and Hollenstein and colleagues (e.g., Granic & Hollenstein, 2003; Granic, Hollenstein, Dishion, & Patterson, 2003; Granic & Lamey, 2002; Hollenstein et al., 2004). The extension of traditional models based on conditional probabilities to allow for path dependence is an exciting open area of research that could use various new developments in complex systems. Gardner (1990) presented a Markov model for sequential categorical data that allows for heterogeneity. W. Griffin and Gardner (1989) provided a framework for path independent models of duration data in social interaction and pointed to the difficulty of adding heterogeneity to such models within a classical statistical approach.

CONCLUSION

In this chapter, I have presented a brief conversation about interesting issues that a transactional perspective to developmental psychology poses for statistical models. There are many details I did not elaborate and many issues I could not discuss. Sameroff and MacKenzie (2003), for example, discussed examples of the transactional model involving experimental manipulations, quasi-experimental designs, naturalistic observations, and intervention studies. There have been many exciting developments in the statistics community around the problems of inferring causality in cases in which randomization is not possible (e.g., Heckman, 2005; Rubin, 2005). The general approach is to model the observation and its corresponding counterfactual, such as what the observation would have been had the case been in another condition. This approach has provided several new advances: new ways of dealing with covariates such as propensity scores, new ways of dealing with differential dropout rates, new ways of handling matching designs, and new ways of modeling missing data. It may provide a useful framework for developing some of the extensions that are needed to implement the transactional model.

There has been a long-standing appreciation of the power of models in our understanding of developmental processes (Sameroff & Chandler, 1975; Sameroff & MacKenzie, 2003; Sigel & Parke, 1987). Likewise, there are many new developments in the statistics and biostatistics literature for general models of repeated measurements that have direct relevance to the transactional model (e.g., Lindsay, 2004; Vonesh & Chinchilli, 1997). But there is much left to develop. This chapter has met its goal if it has alerted developmental psychologists to some of these new statistical tools and has made it clear that for these statistical tools to be useful one needs to be clear about the features of the psychological model. There is a transactional model of sorts between the theory one wants to test and the statistical procedures one uses to test that theory.

REFERENCES

Barnett, R. C., Marshall, N. L., Raudenbush, S. W., & Brennan, R. T. (1993). Gender and the relationship between job experiences and psychological distress: A study of dual-earner couples. *Journal of Personality and Social Psychology, 64,* 794–806.

Boker, S. M., & McArdle, J. J. (1995). Statistical vector field analysis applied to mixed cross-sectional and longitudinal data. *Experimental Aging Research, 21,* 77–93.

Boker, S. M., & Nesselroade, J. R. (2002). A method for modeling the intrinsic dynamics of intraindividual variability: Recovering the parameters of simulated oscillators in multi-wave panel data. *Biometrika, 37,* 127–60.

Busey, T. A., & Loftus, G. R. (1994). Sensory and cognitive components of visual information acquisition. *Psychological Review, 101,* 446–469.

Cook, W., & Kenny, D. (2004). Application of the social relations model to family assessment. *Journal of Family Psychology, 18,* 361–71.

Cook, W., & Kenny, D. (2005). The actor–partner interdependence model: A model of bidirectional effects in developmental studies. *International Journal of Behavioral Development, 29,* 101–109.

Cudeck, R., & du Toit, S. H. C. (2002). A version of quadratic regression with interpretable parameters. *Multivariate Behavioral Research, 37,* 501–519.

Durlauf, S. N., & Young, H. P. (2001). *Social dynamics.* Brookings Institution Press.

Erev, I., & Roth, A. E. (1998). Predicting how people play games: Reinforcement learning in experimental games with unique, mixed strategy equilibria. *American Economic Review, 88,* 848–881.

Gardner, W. (1990). Analyzing sequential categorical data: Individual variation in Markov chains. *Psychometrika, 55,* 263–275.

Gelman, A., & Hill, J. (2007). *Data analysis using regression and multilevel/hierarchical models*. New York: Cambridge University Press.

Gonzalez, R., & Griffin, D. (1999). The correlational analysis of dyad-level data in the distinguishable case. *Personal Relationships, 6*, 449–469.

Gonzalez, R., & Griffin, D. (2000). The statistics of interdependence: Treating dyadic data with respect. In W. Ickes & S. W. Duck (Eds.), *The social psychology of personal relationships* (pp. 181–214). Chichester, England: Wiley.

Gonzalez, R., & Griffin, D. (2001). A statistical framework for modeling homogeneity and interdependence in groups. In G. Fletcher & M. Clark (Eds.), *Handbook of social psychology: Vol. 2. Interpersonal processes* (p. 505–534). Malden, MA: Blackwell.

Gonzalez, R., & Griffin, D. (2002). Modeling the personality of dyads and groups. *Journal of Personality, 70*, 901–924.

Gonzalez, R., & Wu, G. (1999). On the shape of the probability weighting function. *Cognitive Psychology, 38*, 129–166.

Gottman, J. M., Murray, J. D., Swanson, C., Tyson, R., & Swanson, K. R. (2002). *The mathematics of marriage: Dynamic nonlinear models*. Boston: MIT Press.

Granic, I., & Hollenstein, T. (2003). Dynamic systems methods for models of developmental psychopathology. *Development and Psychopathology, 15*, 641–69.

Granic, I., Hollenstein, T., Dishion, T. J., & Patterson, G. R. (2003). Longitudinal analysis of flexibility and reorganization in early adolescence: A dynamic systems study of family interactions. *Developmental Psychology, 39*, 606–617.

Granic, I., & Lamey, A. V. (2002). Combining dynamic systems and multivariate analyses to compare the mother–child interactions of externalizing subtypes. *Journal of Abnormal Child Psychology, 30*, 265–283.

Griffin, D., & Gonzalez, R. (1995). The correlational analysis of dyad-level data: Models for the exchangeable case. *Psychological Bulletin, 118*, 430–439.

Griffin, W. (2000). A conceptual and graphical method for converging multisubject behavioral observational data into a single process indicator. *Behavioral Research Methods, Instruments, & Computers, 32*, 120–133.

Griffin, W., & Gardner, W. (1989). Analysis of behavioral durations in observational studies of social interaction. *Psychological Bulletin, 106*, 497–502.

Haavelmo, T. (1943). The statistical importance of a system of simultaneous equations. *Econometrica, 11*, 1–12.

Heckman, J. (2005). The scientific theory of causality. *Sociological Methodology, 35*, 1–78.

Hollenstein, T., Granic, I., Stoolmiller, M., & Snyder, J. (2004). Rigidity in parent–child interactions and the development of externalizing and internalizing behavior of early childhood. *Journal of Abnormal Child Psychology, 32*, 595–607.

Ickes, B., & Gonzalez, R. (1994). Social cognition and social cognition: From the subjective to the intersubjective. *Small Group Research, 25*, 294–315.

Kenny, D. A. (1979). *Correlation and causality*. New York: Wiley.

Kenny, D. A., Kashy, D. A., & Cook, W. L. (2006). *Dyadic data analysis*. New York: Guilford Press.

Kenny, D. A., & la Voie, L. (1985). Separating individual and group effects. *Journal of Personality and Social Psychology, 48*, 339–48.

Kim, J. G., Menzefricke, U., & Feinberg, F. M. (2007). Capturing flexible heterogeneous utility curves: A Bayesian spline approach. *Management Science, 53*, 340–354.

Lindsay, J. K. (2004). *Statistical analysis of stochastic processes in time*. Cambridge, England: Cambridge University Press.

Nagin, D. S. (1999). Analyzing developmental trajectories: A semiparametric, group-based approach. *Psychological Methods, 4*, 139–157.

Nowak, A., & Vallacher, R. R. (1998). *Dynamic social psychology*. New York: Guilford Press.

Olsen, J. A., & Kenny, D. A. (2006). Structural equation modeling with interchangeable dyads. *Psychological Methods, 11*, 127–141.

Pitt, M. A., Kim, W., Navarro, D. J., & Myung, J. I. (2006). Global model analysis by parameter space partitioning. *Psychological Review, 113*, 57–83.

Rogosa, D., Brandt, D., & Zimowski, M. (1983). A growth curve approach to the measurement of change. *Psychological Bulletin, 92*, 726–748.

Rouder, J. N., Lu, J., Speckman, P., Sun, D., & Jiang, Y. (2005). A hierarchical model for estimating response time distributions. *Psychonomic Bulletin & Review, 12*, 195–223.

Rubin, D. (2005). Causal inference using potential outcomes: Design, modeling, decisions. *Journal of the American Statistical Association, 100*, 322–331.

Salmon, T. (2001). An evaluation of econometric models of adaptive learning. *Econometrica, 69*, 1597–1628.

Sameroff, A. (1995). General systems theories and developmental psychopathology. In D. Cicchetti & D. J. Cohen (Eds.), *Developmental psychopathology: Vol. 1. Theory and methods* (pp. 659–695). New York: Wiley.

Sameroff, A. (2000). Developmental systems and psychopathology. *Development and Psychopathology, 12*, 297–312.

Sameroff, A., & Chandler, M. (1975). Reproductive risk and the continuum of caretaking casualty. *Review of Child Development Research, 4*, 187–244.

Sameroff, A., & MacKenzie, M. J. (2003). Research strategies for capturing transactional models of development: The limits of the possible. *Development and Psychopathology, 15*, 613–640.

Savit, R., Koelle, K., Treynor, W., & Gonzalez, R. (2004). Man and superman: Human limitations, innovation, and emergence in resource competition. In K. Tumer & D. Wolpert (Eds.), *Collectives and the design of complex systems* (p. 199–212). New York: Springer Publishing Company.

Sigel, I. E., & Parke, R. D. (1987). Structural analysis of parent–child research models. *Journal of Applied Developmental Psychology, 8*, 123–137.

Sutton, R. S., & Barto, A. G. (1998). *Reinforcement learning: An introduction*. Boston: MIT Press.

Tversky, A. (1977). Features of similarity. *Psychological Review, 84*, 327–352.

Tversky, A., & Gati, I. (1982). Similarity, separability, and the triangle inequality. *Psychological Review, 89*, 123–154.

Vonesh, E. F., & Chinchilli, V. M. (1997). *Linear and nonlinear models for the analysis of repeated measurements*. New York: Marcel Dekker.

Woody, E., & Sadler, P. (2005). Structural equation models for interchangeable dyads: Being the same makes a difference. *Psychological Methods, 10*, 139–158.

Wu, G., & Gonzalez, R. (1996). Curvature of the probability weighting function. *Management Science, 42*, 1676–1690.

Yang, S., Chen, Y., & Allenby, G. (2003). Bayesian analysis of simultaneous demand and supply. *Quantitative Marketing and Economics, 1*, 251–75.

13

PURSUING A DIALECTICAL PERSPECTIVE ON TRANSACTION: A SOCIAL RELATIONAL THEORY OF MICRO FAMILY PROCESSES

LEON KUCZYNSKI AND C. MELANIE PARKIN

It is well recognized that the transactional model proposed by Sameroff (1975a, 1975b) is a model of qualitative change. Sameroff asserted that the transactional model concerned qualitative rather than incremental change and that the underlying process was dialectical rather mechanistic in nature. Parents and children were proposed to be engaged in continual transformation as each partner reinterprets and responds to new emerging characteristics of the other. What is less recognized is that to be implemented, the model also requires a qualitative change on the part of researchers. Concepts such as transformation, dialectics, and the central role of representations as the locus of change were all revolutionary ideas at a time when the norm was research that focused on continuity, unidirectional influence, and mechanistic conceptions of process.

How far has this revolution progressed? Despite the importance of the transactional model for consolidating the idea of bidirectional influence, few researchers have taken up the challenge of Sameroff's dialectical conception of transactional processes (Sameroff & MacKenzie, 2003, p. 619). Most researchers have instead adopted a behavioral model of transaction that conceives of

This research was supported by a grant to Leon Kuczynski from the Social Sciences and Humanities Research Council of Canada.

bidirectionality in terms of reciprocal exchanges of behaviors. Behavioral or mechanistic models that consider phenomena in terms of individual decontextualized behaviors, passive reactivity, continuity, and linear outcomes continue to be preferred over organismic–systemic metaphors that emphasize nonlinear change. Further, questions regarding the products of socialization continue to be more compelling than questions regarding the underlying processes.

The purpose of this chapter is to explore what it would take to meaningfully implement Sameroff's (1975b) transactional model. Although Sameroff asserted a dialectical model, he did not fully explicate its assumptions about parents and children and how they relate to each other during interactions. In our view, the dialectical model of transaction was ahead of its time and required a framework of supportive theoretical models and knowledge regarding parent–child relationships as well as a shift in the attitudes and orientations of researchers. To illustrate this view, we begin by presenting the slow personal steps toward acceptance of the transactional model from the perspective of Leon Kuczynski, a long-term socialization researcher who was a graduate student at the inception of the model. We document his journey to acceptance of transactional models through a process of self-discovery of a number of assumptions that we believe are prerequisites for full acceptance of the model. Next, we outline steps that need to be taken by researchers who would like to implement a transactional model in their research. The chapter ends with a reflection by C. Melanie Parkin, a doctoral student who is grappling with the transactional model in the current day.

As a behaviorally trained graduate student in the 1970s, I (Kuczynski) recall that I was not ready for the idea of transaction. I did not comprehend the concept of dialectics but was attracted to the figure of reciprocal bidirectional arrows connecting the parent and child over time in Sameroff's initial article (1975a). Thus, as with many other researchers, mine was a case of accepting the model figuratively but not literally. Despite this, my interest in the processes underlying measured variables set me on a path that ultimately led back to a dialectical conception of transaction. Particularly important was that my research favored the theory discovery rather than the theory-testing mode of the methodological cycle (Kuczynski & Daly, 2003). Methods such as observing natural parent–child interactions and open-ended interviews regarding parental behavior in different social contexts continually provided data that were at odds with the dominant research measures and conceptions in the field.

My early research experiences convinced me of the inadequacy of measurement approaches that attempted to capture complex parental and child processes in terms of stable characteristics. I never could get the darn stable trait things to work! The first lesson was a series of studies (Dawber & Kuczynski, 1999; Grusec & Kuczynski, 1980; Kuczynski 1984; Trickett & Kuczynski, 1987) indicating that parents' use of discipline strategies such as reasoning and power assertion was highly situation dependent. When parents were confronted with different misbehaviors, they interpreted them and adapted their control strategies in qualitatively different ways depending on their interpretation. Where the then-dominant measures focused on parental traits and styles, these observations revealed that parents interpret situations and behave strategically to meet their goals. A second line of research, this time focused on the construct of children's noncompliance, taught a similar lesson. Children's noncompliance takes qualitatively different forms: defiance, passive noncompliance, polite refusal, compromise, and negotiation. Moreover, children use these as persuasive strategies to influence parents as a way of achieving their goals (Kuczynski, 1984; Kuczynski & Hildebrandt, 1997; Kuczynski & Kochanska, 1990; Kuczynski, Kochanska, Radke-Yarrow, & Girnius-Brown, 1987).

These two empirical discoveries led to a confrontation with theory. Although there were many similar empirical studies of purposeful and adaptive microprocesses at play, there were no overarching theoretical frameworks in the socialization literature that assumed that parents and children interacted strategically with each other. This led to an interlude of literature review and model building during which my colleagues and I attempted to identify a coherent set of theoretical assumptions that seemed to underlie such studies.

First, our findings on the adaptive nature of parent and child behaviors implicated a role for human agency, the capacity to construct meanings and act on them. Second, our reviews of the literature on parent–child interactions since the 1970s revealed many phenomena, including the fact that parents tolerate noncompliance, conflict, and demands from their own children that they would not from strangers. This required another assumption that accounted for the very existence of bidirectional processes. Namely, that parents and children are not isolated individuals but are part of a larger long-term relationship system that makes possible the enhanced bidirectionality one sees in parent–child relationships (Kuczynski & Hildebrandt, 1997; Kuczynski, Marshall, & Schell, 1997; Lollis & Kuczynski 1997). The last piece was the realization that once we had interacting agents and bidirectional influences and were considering parents and children as an interdependent system, we were moving from a generic model of bidirectionality to a specific dialectical one (Kuczynski & Navara, 2006; Kuczynski & Parkin, 2006). And that is how I came full circle from rejecting the transactional model to embracing it in the space of 30 years!

SOCIAL RELATIONAL THEORY: FIVE STEPS TOWARD A TRANSACTIONAL MODEL

Kuczynski and Parkin (2007) proposed a social relational theory of socialization as a microprocess counterpart to the transactional model. In so doing, we argued, as did Patterson (1997), that a complete research program on socialization needed both a macroprocess predictive model that examined changes and continuities in measures of child or parent over time as well as a theoretically compatible microprocess model to explore underlying dynamics of social interaction. Social relational theory builds on a bilateral framework of interconnected assumptions regarding parent–child relationships (Kuczynski, 2003). These include (a) a dialectical model of bidirectional causality, (b) parents and children as human agents, (c) relationship as a context for interaction, and (d) interdependent asymmetry in power. In the next section we address the following question: What would it take for a researcher to take the transforming leap to a transactional model of development based on dialectics? We outline five steps that we believe would assist such a qualitative change in the orientation of researchers. None of the steps are novel. Rather, like the concept of bidirectionality itself, most researchers acknowledge the majority of these ideas without meaningfully implementing them in their research questions and designs. Therefore, as an encouragement to personal action, we also prescribe a corresponding series of five personal resolutions for researchers who would like to attempt a transforming leap into transactional research.

Step 1: A Dialectical Model of Bidirectionality

Dialectics, as it is applied to the transactional model, is a basic model of dynamic systems. As such, it is a metatheory about the inherent systemic nature of all phenomena and of the nature of change, namely, that everything and every process consists of opposing forces that actively relate through a process of contradiction, thereby producing unending qualitative change. This dialectical conception of causality stands in contrast to behavioral conceptions of bidirectionality that consider social interactions in terms of an interconnected series of behavioral stimulus–response sequences (Sears, 1951), reciprocal chains of negative reinforcement (Patterson, 1982), or homeostatic control systems (Bell & Harper, 1977), which were proposed as mechanisms of incremental shaping of behaviors over time.

Valsiner (1989) formally described the dialectical process as follows:

> The relation between X and Y is *contradictory* (a basic assumption of the dialectical perspective) in the sense that the two parts (X and Y) are opposing each other while remaining mutually necessary parts of the system. As a result of the opposition of the subparts of the whole, the whole system "leaps" to a novel state of being (incorporating a new part Z). (p. 67)

Three features of this definition need to be examined more closely because they reinforce the idea that dialectical theory is a package deal involving a number of interconnected assumptions. Dialectics assumes that X and Y are not isolated individuals but parts of a larger interconnected system. The model assumes that contradiction, rather than shaping or stimulus–response reactivity, is the basic process for change. Last, it assumes that the expected outcome of a dialectical process is *novelty* rather than continuity. The *new part* Z is also known as the *transaction*, *new synthesis*, *turning point*, *phase shift*, *changed representation*, or *new working model*. An example to be examined at length in the section that follows is a child entering adolescence and becoming more autonomy seeking (X), conflicting with previous parent restrictions (Y), and producing a new accommodation (Z).

Resolution 1: Challenge Linear Thinking

Although processes such as reactivity and incremental shaping have their place in understanding human behavior, researchers interested in qualitative change need to grapple with dialectical outcomes and processes. The idea of dialectics alerts researchers that the outcomes of socialization processes must be more than exact compliance or the mere transmission of similarity from the older generation to the younger generation. Although people may long for linear outcomes, much as they prefer stories to have "happy endings" rather than uncertain dialectical ones (such as the blank screen at the end of the *Sopranos* television series), continuous change and new syntheses are also to be expected outcomes of socialization and development. A parent's strategies can produce a range of unintended responses beyond compliance or intergenerational transmission of values. For instance, children may resist or creatively interpret parental messages; they may evade parental influence; or their responses may cause parents to change. Thus socialization is better understood as producing a range of positive or negative trajectories rather than precisely determined outcomes.

Two potential barriers need to be confronted before researchers can leave their incremental journey and leap to a dialectical model of causality. First, perhaps a dialectical view of causality is too complex and unintuitive to be useful for thinking about causality. This view is contradicted by a popular Web site, Dialectics for Kids (http://home.igc.org/~venceremos/index.htm). Its premise is that dialectical concepts are such common sense for understanding basic physical, biological, and psychological processes that even children can quickly grasp the ideas.

> Dialectics is a tool to understand the way things are and the way things change. Understanding dialectics is as easy as 1–2–3. One—Every thing (every object and every process) is made of opposing forces/opposing sides. Two—Gradual changes lead to turning points, where one opposite overcomes the other. Three—Change moves in spirals, not circles. (What the Heck is Dialectics?, n.d., ¶1–¶4)

The Web site goes on to provide illustrations intended for ages 4 through adulthood that demonstrate how people's everyday understanding of objects and processes involves dialectical thinking.

A second potential barrier is the perception that dialectics is too philosophical to be useful for science. Glassman (2000) suggested that dialectical ideas have in fact been appropriated by many theories in developmental psychology, including cognitive development, ecological and dynamic systems theory, and attachment theory, but the dialectical origins of the ideas are rarely acknowledged. This is because the term originates from philosophers such as Hegel, Feuerbach, Marx, Lévy-Bruhl, Vygotsky, and Riegel and not from empirical research. It is also possible that there is an active avoidance of the term *dialectic* by those who confuse the philosophical idea of dialectical materialism, developed by Marx and his partner Engels, with its failed application to economics and political communism. Researchers who resist philosophy could consider that even behaviorism and mechanism are philosophical positions (Overton, 2006).

Step 2: Considering Parents and Children as Active Agents

Dialectics, as applied to humans, considers parents and children to be active agents and not passive reactors. Baxter and Montgomery (1996) used the term *praxis* for the assumption that people are simultaneously proactive actors and reactive objects within their social worlds. Sameroff's (1975b) treatment of transactional processes is built on an implicit model of human agency in which parents and children interact as agents who create meaning out of their experiences. Sameroff discussed dialectical processes involving an interaction between contradictory meaning systems as the underlying mechanism of the transactional model. "In every developing system, contradictions are generated and it is these contradictions which provide the motivation which lead the organisms to the higher level of organization found in developmental series" (Sameroff, 1975b, p. 74). The capacity to interpret and construct meaning is only one dimension of human agency. A comprehensive treatment of agency would also consider behavioral, cognitive, and motivational dimensions (Bandura, 2001; Cummings & Schermerhorn, 2003; Kuczynski, 2003). Parents and children have capacities to initiate purposeful behavior and to strategically choose methods for influencing each other's behavior. They have the capacity to reflect on their behavior and interpret messages communicated during interaction. They also have motives to assert themselves, resist demands that block their goals, or contravene their self-constructed understanding of social situations.

Accounts of interaction that consider parents and children as agents lead to an alternative conception of bidirectional influence from behavioral models of bidirectionality. In sociocognitive perspectives, bidirectional influence comes about at a cognitive or representational level as parents and children interpret or construct meanings from each other's behaviors and anticipate and accommodate each other's perspectives during interaction (Dix & Branca,

2003; Grusec, Goodnow, & Kuczynski, 2000; Holden & Hawk, 2003). Such models generally conceive of change as resulting from an active process of mutual adaptation brought about by changing cognitive representations of the interaction or the relationship (Kuczynski, Lollis, & Koguchi, 2003).

Resolution 2: Stop Thinking of Parents and Children Solely in Terms of Mindless Behavioral Reactivity or Sets of Observable Variables

The dialectical researcher is curious about the processes that underlie measured variables and, in particular, what parents and children actually do, think, and feel as they engage in interactions with each other; the goals they form prior to interacting; how they construe their experiences; and, especially, how they choose to accommodate each others' ideas or goals in their own lives.

Step 3: Consider the Parent–Child Relationship as a Context for Interaction

The dialectical concept *unity of opposites* implies that individual elements must be understood as interrelated parts of a whole system. An account of development or change that focuses only on individual actions or characteristics, without reference to interpersonal relationships, groups, societies or culture, is necessarily inadequate (Ho, 1998, p. 8). This point has been stated formally by Valsiner and Connolly (2003):

> Development and the appearance of new emergent properties cannot arise in a single component or solely in the context; two or more components are always necessary. Hence, classical notions of linear causality (e.g., cause *X* leads to outcome *Y*) are inappropriate in the case of developing systems. Instead, developmental science operates with the notion of systemic causality (Valsiner, 2000). An outcome is a result of the systemic relationship between parts of a system: system {*a-R-b*}, where *a* and *b* are parts and *R* is their relation, leads to outcome Y. (p. xi)

The major implication of a dialectical systemic view is that not only outcomes but also dynamics underlying social interaction arise from the particular structure and history of the system under consideration. In bilateral relational perspectives, processes having to do with bidirectional influence and the practice and experience of agency and power depend on the relationship context (Kuczynski, 2003; Kuczynski & Parkin, 2007). For example, in considering the relationship system as a context for transactions, one needs to distinguish mother–child from father–child relationships and be cognizant that relationship dynamics may vary depending on cultural norms (Kuczynski & Parkin, 2007; Trommsdorff & Kornadt, 2003).

In addition, dialectics entails a complex understanding of the nature of the system because it assumes that systems contain both shared and contradictory elements that coexist side by side and continually interact. For example, parents and children continually come into conflict, but they are united by the shared bonds of their mutual relationship (Kuczynski & Hildebrandt, 1997). The interplay between interdependence coexisting with conflict produces an ongoing tension between the parent and the child that generates mutual accommodation, adaptation, and negotiation within an interdependent relationship context. Children's agency and effectiveness as agents is enabled by the relationship, and for both parent and child, the competent expression of agency involves accommodating to the mutual constraints of a reciprocal relationship (Kuczynski & Hildebrandt, 1997).

This view of the relationship system is influenced by Hinde (1979, 1997) who proposed that relationships are formed as dyads, accumulate a history of interactions over time, and produce an emergent relationship that creates a context for future interactions. The relationship itself is a cognitive representation of the multitude of interactions that objectively occurred in the history of the relationship. Each partner in the dyad makes sense of the other's behavior through psychological processes such as perception and interpretation to create meanings and expectancies. These meanings and expectancies become consolidated in representations of the relationship, which then form the filter through which parent and child behaviors are experienced (Hinde, 1997; Lollis, 2003). Children and parents do not act merely on the immediate contingencies found in each other's behavior. They are also influenced by the cognitive representations (expectancies) developed from a long history of interactions (Lollis, 2003; Lollis & Kuczynski 1997) as well as representations of the anticipated future of the relationship (goals).

Relationship cognitions are readily apparent in research designs that compare a parent or a child's reactions to people who vary in terms of their relationship with them, for example, their own parent, the parent of a friend, or a teacher. Parents and children use different strategies in different relationships and justify their behaviors by referring to their knowledge of personality and their predictions of future behavioral or cognitive reactions on the basis of how interactions have progressed in the past (Dawber & Kuczynski, 1999; Hildebrandt & Kuczynski, 1998; Kuczynski, Lollis, Parkin, & Oliphant, 2007).

Interdependence is another feature of close relationships that affects the dynamics of social interactions (Kelley et al., 1983). The interdependence of parents and children is a source both of mutual receptivity and of mutual vulnerability because parents and children behaviorally and psychologically depend on each other to have important needs met. Each parent's or child's sense of efficacy and power in the relationship is contingent on his or her ability to influence the other (Cummings & Schermerhorn, 2003; Emerson, 1962). The construct of psychological mattering, construing oneself as being affectively and cognitively significant to the other, has recently been explored as an important dimension of adolescents' and parents' perceptions of themselves in their relationships (Marshall, 2001; Marshall & Lambert, 2006).

Resolution 3: Stop Thinking About Parents and Children as Individuals or Even Individuals Engaged in Interaction

Parent–child interactions need to be considered as embedded in a larger system that minimally consists of the parent–child relationship itself. A dialectical conception of process would consider change as an outcome of tensions between parents and children who have unique personalities, needs, and perspectives but who nevertheless share a close relationship. The concept of the parent–child relationship as a context for socialization raises a host of basic questions about which there is surprisingly little information. Among these are questions about the structure of parent–child relationships, particularly those features of relationships that have implications for parent–child dynamics. Examples include the nature of past expectancies, future goals, and the dynamics of interdependence (Lollis, 2003; Lollis & Kuczynski, 1997). Such analyses may be extended further by the study of differences between mother–child and father–child relationships and distinct relationships between different parent and child dyads as well as cultural differences in norms for parent–child relationships. The principal lesson that will be learned is that there is more to the whole relationship than the sum of its parts.

Step 4: Considering Contradiction as a Basic Process

Contradiction is a process that explains how parents and children, considered as agents, interpret experiences during everyday interactions and across development. An important step in a dialectical analysis is the identification of contradictions that are inherent or implicit in the phenomenon under consideration. Contradictions can occur within the individual, between individuals, and between individuals and biological and ecological contexts (Reigel, 1976). Parent–child relationships constantly create both external and internal contradictions that feed into the dialectical process. Parenting inherently involves constant adaptation to a rapidly changing organism (Holden & Ritchie, 1988), and parental responsivity to changes in children is the hallmark of parental competence. Children must also constantly adapt to long-term and short-term changes in their parents. Each parent–child relationship can be conceived as a dynamically interacting dyad whose members mutually and continuously influence each other over time (Collins & Madsen, 2003; Lollis & Kuczynski 1997). There are simultaneous developmental and contextual changes for children as they develop from infancy to adulthood and for parents as they move through young adulthood to old age. The intertwining of the lives of parents and children provides constant opportunities for each to experience contradictions in their relationships and primes the system for turning points and qualitative change. Three types of contradiction, *conflict*, *ambivalence*, and *expectancy violations*, are universal forms of tension in parent–child relationships and are suitable foci for the analysis of dialectical processes.

Conflict

In the 1980s, a turning point occurred in research on conflict, from a view that equated conflict with deviant noncompliance on the part of the child to a view of conflict as an inevitable and mutually tolerated aspect of living in close relationships (Dunn & Munn, 1985; Shantz, 1987). Interactions in which parents make demands of their children often pit the parent's needs, goals, will, and interpretations of events against those of the child. In addition, parents and children continuously are exposed to a larger ecological system of values and understandings, including those of peers, television, Internet, and school, that offers up contradictions and opportunities for further change as parents and children adapt to each others' choices (Kuczynski et al., 1997).

The relational view of conflict also introduced dialectics as an alternative to a deterministic view of causality in parent–child relationships and socialization. In a relational view, conflict is viewed as a necessary condition for change, for better or for worse (Shantz & Hartup, 1992). Research from this perspective found outcomes of parent–child conflict to be inadequately captured by deterministic concepts such as compliance and noncompliance. Instead it is more accurate to conceive of conflict as generating temporary novel syntheses for the parent–child dyad, including accommodation (Kuczynski & Hildebrandt, 1997) negotiation, compromise, withdrawal from conflict without resolution, mutual standoff (Eisenberg, 1992; Vulchinch, 1987), and relationship repair following interactional missteps by the parent or child (Harach & Kuczynski, 2005).

Ambivalence

Recently, ambivalence has emerged as a concept for exploring complex psychological processes having to do with individuals' frequent experience of events, emotions, and meanings that simultaneously pull them in different directions. In sociocultural perspectives, ambivalence is a recurring phenomenon of daily experience that continually offers the possibility of creating new meaning and launching individuals on novel trajectories (Abbey & Valsiner, 2004; Valsiner, 2006).

Lüscher and Pillemer (1998) proposed a theory of intergenerational ambivalence for parent–child relations in later life as a tool for understanding how individuals experience and manage contradictory impulses and perceptions in their close relationships. According to Lüscher (2004), intergenerational ambivalence occurs "when polarized simultaneous emotions, thoughts, violations, actions, social relations, and/or structures that are considered relevant for the constitution of individual or collective identities are (or can be) interpreted as temporarily or even permanently irreconcilable" (p. 36). The phenomenon of ambivalence has been uncovered by qualitative researchers who noted the simultaneous presence of positive and negative assessments of intergenerational close relationships including mixed feelings such as simultaneous affection and irritation—a reality that is obscured in ratings of "somewhat close" on forced-choice global assessments of relationships. Although research on ambivalence is in its beginning stages, it is clear that it is a frequent phenomenon in intimate relationships. In a study of 187 participants ranging in age from 13 to 99, Fingerman and Hay (2004) found ambivalence to be more characteristic of relationships involving romantic partners, mothers, fathers, sons and daughters, and siblings than those with extended family or friends.

The parental role involves diverse conflicting functions such as providing authority, caregiving, and security, while at the same time seeking intimacy and long-term connection with children (Harach & Kuczynski, 2005). Acting on behalf of the child is itself a daily exercise of internal contradiction and ambiguity (Holden & Ritchie, 1988). For example, parents must allow exploration yet guard against danger, seek obedience yet allow autonomy, be involved yet not over involved, promote autonomy yet maintain interdependence. There are ample opportunities for ambivalence as parents must act in children's behalf without knowing for certain what will work and how children will evaluate or question their choices at a later date (Lüscher, 2004).

Children also experience ambivalence in their relationships with parents. Ambivalence has been long been inferred as a contributor to insecure attachment (Maio, Fincham, Camillo, & Paleari, 2004) during infancy and is apparent in assessments of mothers and fathers in adolescence (Youniss & Smollar, 1985). In a recent study of Vietnamese youth in Canada (Le, Kuczynski, & Boiger, 2007), themes of ambivalence were found at many levels. For instance, youth may dislike their parents' traditional values but remain protective of their relationships with them; they may agree with some values but disapprove of the means that parents used to achieve them; they may decry the lack of closeness of their relationship with their parents but vow to support their parents in old age.

In a recent study (Parkin & Lataigne, 2007), adolescents talked about managing the information they gave to their parents as a way of protecting the relationship from the inevitable clash of values that would ensue if they were to raise issues openly. For example, one 18-year-old man spoke about his parents' rigid expectations regarding academics and his own partial acceptance of their values:

> Like I know where their [*sic*] coming from, like they believe with a higher education, and with higher marks you'll get further in life. I do understand that right, but I do also know there are other options besides getting a higher education, like you can go into apprenticeship and stuff right? So I'll always keep that open. (Parkin & Lataigne, 2007)

He also explicitly talked about strategies he used to manage his parents' expectations for constant studying and his own goals of maintaining a positive relationship while pursuing his own interests:

> I make sure that they actually see me studying during the day, and then I'll be like, oh I've been studying all day, I'm just going out to take a break. . . . so I'll always have music on, and every time they come around I'd make sure they would see me studying, plus I won't go out every night, I'll space it every other night, every 2 or 3 nights. (Parkin & Lataigne, 2007)

Valsiner (2006) asserted that nonambivalence is a temporary condition in which overcoming one state of ambivalence sets the stage for a new state of ambivalence. An important question for research on parent–child relationships is how ambivalence can be managed in daily life to keep open, and potentially damaging, conflict between generations from arising. However, from a dialectical standpoint, individuals' responses to ambivalence introduce the potential for generating novelty. Choosing to ignore contradictory information, avoiding communication about contentious topics, communicating to arrive at new solutions, or going into therapy all may involve qualitative changes in behaving, relating with others, or perceiving situations.

Expectancy Violations

Receiving information that violates habitual ways of perceiving behavior or persons is a third source of contradiction. Children's transgressions against parental expectations for behavior and parental transgressions against children's expectations create internal contradictions that parents and children must reconcile or adapt to. Sameroff (1975b) provided the first example of expectancy violations as a form of opposition that leads to qualitative change:

> The contradiction that has occurred consists between a meaning system which sees the child as an object to be manipulated, and one which sees the child as a center of needs and desires existing independently of the needs and desires of his parents. . . . The dialectical model would posit at each stage the contradictions with which the mother is faced in trying to understand her child. (p.77)

The expectancy-violation realignment model (Collins, 1995) attempts to explain how families adapt to change as children move through adolescence to adulthood. Developmental change on the part of children creates contradictions in the parents' representation of the child. Although change in the child initially creates conflict, puzzlement, or anguish due to violations in the parent's expectancies, a new basis for interaction will occur when the parent adapts to the new reality by reinterpreting the meaning of the child's behavior and perhaps viewing it as an appropriate sign of autonomy seeking rather than defiance. According to Collins, in early adolescence children tend to increase their demands for autonomy and may be more questioning toward parental authority. Parents may interpret these changes negatively as "attitude" or defiance. The conflict is ultimately resolved when the parent can create a novel meaning of the child's behaviors in terms of signs of emerging adulthood. In dialectical terms, the parent reaches a turning point and begins to respond to the child in a qualitatively different manner because he or she no longer views him- or herself as an authority over the child but more as a consultant or advisor. Adolescents may also undergo transformations in their perceptions of and relationships with parents. Youniss and Smollar (1985) documented how in early adolescence children view parents as unilateral authority figures but eventually come to understand parents as individual personalities with unique strengths and weaknesses.

Resolution 4: Consider Processes Within the Parent and the Child or Between the Parent and Child That Are Most Likely to Be a Source of Contradiction and Disequilibrium

These processes should not be viewed in a negative manner but as inevitable characteristics of human agents living in close relationships. Points of harmony and shared goals are unlikely to be sources of qualitative change (Reigel, 1976). Moreover, considering a process or interaction from the point of view of only one person, such as the parent, provides an overly optimistic view of that person's influence and obscures the conflicts, ambivalences, and violated expectancies of the other.

Step 5: Actively Looking for Points of Qualitative Change

We now move to another element of the transactional model, namely, the locus of the transaction. As stated previously, the early writings of Sameroff (1975a, 1975b) pointed to meaning systems or cognitive representations as the sites of change. Surprisingly, there has been little research or theoretical analysis of this crucial element of the transactional model (Sameroff & MacKenzie, 2003).

How does one identify a transaction? A transaction involves a turning point or qualitative shift in representations, but what does that change look like and where can it be found? There are various ways to conceptualize transactions, and we focus on two that seem useful given the current state of research. First, qualitative change can occur suddenly, or it can occur after a period of gradual, incremental change. Second, qualitative change can occur in the form of microprocess transactions during daily interactions between parents and children or it can occur as a series of transactions during macro process development and life course transitions that may take years for completion.

Catastrophic, Sudden Change

The most salient examples of transactions under a dialectical model are sudden changes in context or knowledge. These are events that are well remembered by those experiencing them and are easiest to recall. A common situation concerns parents' sudden discovery of new information that contradicts their previous understanding of their child. For example, consider the father who first discovers his daughter is sexually active. This knowledge, suddenly gained, may completely contradict his representation of his child, her developmental stage, and his role as a parent. This contradiction may throw the parents' internal working model or representation of the relationship into disarray while the parent comes to terms with the knowledge and develops a new way of thinking about his child. A reversal of this example is a daughter learning that her parents are sexually active and the contradictions that this entails in terms of seeing her parents both in the role of parents and also as individuals with their own goals and desires. By working through this contradiction the child may develop more adequate multifaceted understanding of parents as human beings. These examples all involve the intrusion of new contradictory information that individuals must deal with at a cognitive level.

Sudden change in representations can also be brought about by changes in context. Granic and Patterson (2006) gave the example of a parental divorce. During a parental divorce, a family's context may be completely changed with new living arrangements, standard of living, composition, and dynamics. These changes throw the family system, as well as each person's representations, into a state of perturbation such that old ways of thinking and acting are no longer useful. Under a dialectical framework, one would focus on the ambivalences and expectancy violations that come about with the parents' divorce. For example, children may experience violations of their understanding of their parents' relationship and the nature of marriage and may experience ambivalence about the changes in their lives. It is here that one may usefully look for changed representations signaling a turning point or new synthesis (impermanence of marriage, representations of each parent, redefinitions of family, attributions of causality in which the child may blame him- or herself). In other words, the contradictions brought about by the change in context (divorce) force the child to do more cognitive work regarding evaluating new context, interpreting emotional reactions, readdressing beliefs, and thereby, creating a new, potentially more sophisticated meaning system.

Sudden changes occur not just on a grand scale such as with parental divorce but also on the more mundane scale of moment-to-moment interactions. Parents are called on many times a day to make decisions about how to respond to unexpected or perturbing events or discoveries regarding a child's behavior. These may cause a qualitative change in the parent's strategic interactions with a child. The most well-researched example concerns situational changes in parents' responses to children's transgressions (Grusec & Kuczynski, 1980; Kuczynski, 2003; Trickett & Kuczynski, 1986). The most commonly used parental strategy during everyday interaction is mild power assertion such as unexplained commands, suggestions, and prohibitions that essentially are used to obtain immediate compliance. However, when parents interpret the child's behavior as requiring long-term change, for example, when the transgression involves a safety issue or a moral offence, parents qualitatively change their strategies to a more considered way of interacting and begin to provide reasons, including emotion-laden, other-oriented inductions. Here, a changed context evokes a changed representation of the child's behavior propelling the parent to reflect on the situation and problem solve other ways of interacting. Holden and Hawk (2003) have called this process *metaparenting*.

Gradual Change Leading to Qualitative Change

A classic view of dialectics is that incremental changes can accumulate until they reach a turning point where they precipitate a qualitative change that throws the entire system out of its usual pattern. Examples of changed representations due to incremental change are more difficult to pinpoint than those that occur as a result of sudden qualitative change in context. Many of the examples of transaction cited by Sameroff and MacKenzie (2003), such as studies by Bugental and Shennum (1984); Rubin, Nelson, Hastings, and Asendorpf (1999); and Stern and Hildebrandt (1986), focused on changes in mothers' interpretations of their children's behavior or cognitive representations of relationships as the locus of qualitative changes (see chap. 3, this volume). These studies suggest qualitative change in parental representations over time; however, it may be difficult for individuals to identify the point at which the transformation took place. In some cases, an incrementally changed way of thinking or responding may go unnoticed until it in some way becomes a problem and is brought into awareness. For example, Patterson's (1997) research on aggressive boys and their mothers' habitual ways of responding suggests the possibility that when the aggressive behavior reaches a certain level, mothers are forced to reflect on the behavior and their own reactions, thus bringing a habitual or automatic process into awareness and marking a turning point in the meaning of the phenomenon for better or for worse.

Patterson, Reid, and Dishion, (1992) described a number of phenomena that appear to involve transactions in which incrementally changing patterns of interaction qualitatively change their form in a process they identified as escalation. Examples include insignificant verbally coercive exchanges that at a certain critical frequency escalate to physical aggression, or habitual patterns of coercive behavior in children that escalate to participation in deviant peer groups and antisocial activities. Granic and Patterson (2006) recently implicated cognitive and emotional processes as driving forces in such escalations, but thus far, they are usually inferred rather than studied in their own right.

Resolution 5: Actively Theorize and Search Out Points of Qualitative Change, Particularly Points at Which Parents and Children Change Their Representations of Each Other or of the Meaning of Their Behaviors in Ongoing Interactions

Sameroff and MacKenzie (2003) drew attention to the need for research on the transactional model to provide evidence of qualitative change and discussed a number of statistical and experimental approaches as useful future directions. To this we add the need for naturalistic observational and qualitative research to identify and describe the phenomenon of changed representations and points of transaction. As has happened before in the history of socialization research, it appears that investigators skipped the descriptive and discovery stage of research on transactions and launched prematurely to the theory-testing stage (McCall, 1977). In the current era, much more is known about the process of interaction than about the processes of transaction. What is needed is a theory-driven methodology that elucidates the incidence and characteristics in transactions between parents and children in everyday life. Qualitative methods are especially suitable for this purpose (Kuczynski & Daly, 2003). The analysis of open-ended interviews is the cognitive counterpart of direct observation in that the focus is on the inductive analysis of parents' and children's reports of their naturally occurring representations of their interactions with each other.

The search for transactions may well be guided by the following principles. First, not everything that occurs between parents and children involves human agency. Many interactions occur at an automatic level (Patterson et al., 1992). However, transaction, or qualitative change, happens at the level of representations and involves cognitive and emotional processes. Second, points of qualitative change are best studied during periods of contradiction and transition rather than during times of harmony and stability. As many have pointed out (e.g., Granic & Patterson, 2006; Sameroff, 1975a), researchers have tended to use methods more suitable for uncovering stability rather than change. It may be that the concept of agency is especially needed when one looks at change but not when one looks at stability, which can easily be accounted for by habit and automatic processes. Third, the task of identifying qualitative change is likely to be easiest at points of sudden or catastrophic change. The more difficult phenomena to locate are points of qualitative change in representations after a long period of incremental change. Historically, the task of uncovering taken-for-granted representations lurking in the unconscious has been the territory of therapy rather than research. However there is room for research methodologies that bring into reflective awareness representations, emotions, and assumptions that may underlie both stability and change in social interactions, practices, and trajectories of development.

CONCLUSION

We began with the narrative of how Leon Kuczynski's training in the 1970s left him ill prepared to comprehend and accept the transactional model. We end with a narrative of C. Melanie Parkin, a doctoral student who is grappling with the model currently.

One would think that students are more prepared today to implement a research program that concerns parents and children as agents, relationship systems, and dialectical processes. The current era evokes considerable ambivalence in the new researcher. As a graduate student immersed in the resolutions and steps outlined in this chapter, I (Parkin) feel excited and invigorated by the challenge; moreover, I am aware that qualitative and observational methods are beginning to come into favor once more in developmental psychology. Nevertheless, traditional research questions and methods remain in the ascendance and therefore, it feels very risky to embark on a research career stemming from this orientation.

Perhaps my ambivalence originates in the criticism made by many that transactional models are vague and not amenable to empirical study. I wonder if this criticism stems from a restricted view of science that favors prediction and hypothesis testing rather than the arduous descriptive work needed for understanding underlying processes. As Sameroff (1975b) stated, "Predictive efficiency is not the same as knowledge of the processes by which development takes place" (p. 67). Currently, there is much "talk" about human agency and process, but the climate continues not to favor studying parents and children as if they were agents. What would it take to implement a transactional model based on dialectics? I suspect that real progress will require a transaction in our discipline and qualitative change in the values and orientations of researchers.

REFERENCES

Abbey, E., & Valsiner, J. (2004, December). Emergence of meanings through ambivalence. *FQS: Forum Qualitative Sozialforschung, 6,* Article 23. Retrieved December 3, 2008, from http://www.qualitative-research.net/index.php/fqs/article/view/515

Bandura, A. (2001). Social cognitive theory: An agentic perspective. *Annual Review of Psychology, 52,* 1–26.

Baxter, L. A., & Montgomery, B. M. (1996). *Relating: Dialogues and dialectics.* New York: Guilford Press.

Bell, R. Q., & Harper, L. V. (1977). *Child effects on adults.* Lincoln: University of Nebraska Press.

Bugental, D. B., & Shennum, W. A. (1984). "Difficult" children as elicitors and targets of adult communication patterns: An attributional–behavioral transactional analysis. *Monographs of the Society for Research in Child Development, 49*(1), 1–79.

Collins, W. A. (1995). Relationships and development: Family adaptation to individual change. In S. Shulman (Ed.), *Close relationships and socioemotional development* (pp. 128–154). Norwood NJ: Ablex.

Collins, W. A., & Masden, S. D. (2003). Developmental change in parenting interactions. In L Kuczynski (Ed.), *Handbook of dynamics in parent child relationships* (pp. 49–66). Thousand Oaks, CA: Sage.

Cummings, E. M., & Schermerhorn, A. C. (2003). A developmental perspective on children as agents in the family. In L. Kuczynski (Ed.), *Handbook of dynamics in parent–child relations* (pp. 91–108). Thousand Oaks CA: Sage.

Dawber, T., & Kuczynski, L. (1999). The question of ownness: Influence of relationship context on parental socialization strategies. *Journal of Social and Personal Relationships, 16,* 475–493.

Dix, T., & Branca, S. H. (2003). Parenting as a goal-regulation process. In L. Kuczynski (Ed.), *Handbook of dynamics in parent–child relations* (pp. 167–188). Thousand Oaks CA: Sage.

Dunn, J. F., & Munn, P. (1985). Becoming a family member: Family conflict and the development of social understanding in the second year. *Child Development, 56,* 480–492.

Eisenberg, A. R. (1992). Conflicts between mothers and their young children. *Merrill-Palmer Quarterly, 38,* 21–44.

Emerson, R. M. (1962). Power-dependence relations. *American Sociological Review, 27,* 21–44.

Fingerman, K. L., & Hay, E. (2004). Intergenerational ambivalence in the context of the larger social network. In K. Pillemer & K. Luscher (Eds.), *Intergenerational ambivalences: New perspectives on parent–child relations later in life* (pp. 133–150). Oxford, England: Elsevier Science.

Glassman, M. (2000). Negation through history: Dialectics and human development. *New Ideas in Psychology, 18,* 1–22.

Granic, I., & Patterson, G. R. (2006). Toward a comprehensive model of antisocial development: A dynamic systems approach. *Psychological Review, 113*, 101–131.

Grusec, J. E., Goodnow, J. J., & Kuczynski, L. (2000). New directions in analyses of parenting contributions to children's acquisition of values. *Child Development, 71*, 205–211.

Grusec, J. E., & Kuczynski, L. (1980). Direction of effect in socialization: A comparison of the parent's versus the child's behavior as determinants of disciplinary techniques. *Developmental Psychology, 16*, 1–9.

Harach, L., & Kuczynski, L. (2005). Construction and maintenance of parent–child relationships: Bidirectional contributions from the perspective of parents. *Infant and Child Development, 14*, 327–343.

Hildebrandt, N., & Kuczynski, L. (1998, May 9). *Children's experience of agency in relationships with their own mothers and familiar and unfamiliar adults*. Paper presented at the 10th Biennial Conference on Child Development, Waterloo, Ontario, Canada.

Hinde, R. A. (1979). *Toward understanding relationships*. London: Academic Press.

Hinde, R. A. (1997). *Relationships: A dialectical perspective*. Hove, England: Psychology Press.

Ho, D. Y. F. (1998). Interpersonal relationships and relationship dominance: An analysis based on methodological relationalism. *Asian Journal of Social Psychology, 1*, 1–16.

Holden, G. W., & Hawk, K. H. (2003). Meta-parenting in the journey of child rearing: A cognitive mechanism for change. In L. Kuczynski (Ed.), *Handbook of dynamics in parent–child relations* (pp. 189–210). Thousand Oaks, CA: Sage.

Holden, G. W., & Ritchie, K. L. (1988). Child rearing and the dialectics of parental intelligence. In J. Valsiner (Ed.), *Child development within culturally structured environments: Parental cognition and adult–child interaction* (pp. 30–59). Norwood, NJ: Ablex.

Kelley, H. H., Berscheid, E., Christensen, A., Harvey, J. H., Huston, T. L., Levinger, G., et al. (1983). *Close relationships*. New York: Freeman.

Kuczynski, L. (1984). Socialization goals and mother–child interaction: Strategies for long-term and short-term compliance. *Developmental Psychology, 20*, 1061–1073.

Kuczynski, L. (2003). Beyond bidirectionality: Bilateral conceptual frameworks for understanding dynamics in parent–child relations. In L. Kuczynski (Ed.), *Handbook of dynamics in parent–child relations* (pp. 1–24). Thousand Oaks, CA: Sage.

Kuczynski, L., & Daly, K. (2003). Qualitative methods for inductive (theory-generating) research: Psychological and sociological approaches. In L. Kuczynski (Ed.), *Handbook of dynamics in parent–child relationships* (pp. 372–393). Thousand Oaks, CA: Sage.

Kuczynski, L., & Hildebrandt, N. (1997). Models of conformity and resistance in socialization theory. In J. E. Grusec & L. Kuczynski (Eds.), *Parenting and the internalization of values: A handbook of contemporary theory* (pp. 227–256). New York: Wiley.

Kuczynski, L., & Kochanska, G. (1990). The development of children's noncompliance strategies from toddlerhood to age 5. *Developmental Psychology, 26*, 398–408.

Kuczynski, L., Kochanska, G., Radke-Yarrow, M. & Girnius-Brown, O. (1987). A developmental interpretation of young children's noncompliance. Developmental *Psychology, 23*, 799–806.

Kuczynski, L., Lollis, S., & Koguchi, T. (2003). Reconstructing common sense: Metaphors of bidirectionality in parent–child relations. In L. Kuczynski (Ed.), *Handbook of dynamics in parent–child relations* (pp. 421–438). Thousand Oaks, CA: Sage.

Kuczynski, L., Lollis, S., Parkin, C. M., & Oliphant, A. (2007, March). *Children's choice of mothers and fathers as targets for influence: Using knowledge of distinct relationships*. Poster session presented at the Biennial Meeting of the Society for Research on Child Development, Boston.

Kuczynski, L., Marshall, S., & Schell, K. (1997). Value socialization in a bidirectional context. In J. E. Grusec & L. Kuczynski (Eds.), *Parenting and the internalization of values: A handbook of contemporary theory* (pp. 23–50). New York: Wiley.

Kuczynski, L., & Navara, G. (2006). Sources of change in theories of socialization and internalization. In M. Killen & J. Smetana (Eds.), *Handbook of moral development*. Mahwah, NJ: Erlbaum.

Kuczynski, L., & Parkin, C. M. (2007). Agency and bidirectionality in socialization: Interactions, transactions, and relational dialectics. In J. E. Grusec & P. Hastings (Eds.), *Handbook of socialization* (pp. 259–283). New York: Guilford Press.

Lawrence, J. A., & Valsiner, J. (1993). Conceptual roots of internalization: From transmission to transformation. *Human Development, 36*, 150–167.

Le, A., Kuczynski, L., & Boiger, M. (2007, October). *The acculturation of Vietnamese parent–child relationships*. Poster presented at the On New Shores Conference, Guelph, Ontario, Canada.

Lollis, S. (2003). Conceptualizing the influence of the past and the future in present parent–child relationships. In L. Kuczynski (Ed.), *Handbook of dynamics in parent–child relationships* (pp. 67–89). Thousand Oaks, CA: Sage.

Lollis, S., & Kuczynski, L. (1997). Beyond one hand clapping: Seeing bidirectionality in parent–child relations. *Journal of Social and Personal Relationships, 14*, 441–461.

Lüscher, K. (2004). Conceptualizing and uncovering intergenerational ambivalence. In K. Pillemer & K. Lüscher (Eds.), *Contemporary perspectives in family research: Vol. 4. Intergenerational ambivalences: New perspectives on parent–child relations in later life* (pp. 23–62). New York: Elsevier.

Lüscher, K., & Pillemer, K. (1998). Intergenerational ambivalence: A new approach to the study of parent–child relations in later life. *Journal of Marriage and the Family, 60*, 413–425.

Maio, G. R., Fincham, F. D., Camillo, R., & Paleari, F. G. (2004). Ambivalence and attachment in family relationships. In K. Pillemer & K. Lüscher (Eds.), *Contemporary perspectives in family research: Vol. 4. Intergenerational ambivalences: New perspectives on parent–child relations in later life* (pp. 285–312). New York: Elsevier.

Marshall, S. K. (2001). Do I matter? Construct validation of adolescents' perceived mattering to parents and friends. *Journal of Adolescence, 24,* 473–490.

Marshall, S. K., & Lambert, J. D. (2006). Parental mattering: A qualitative inquiry into the tendency to evaluate the self as significant to one's children. *Journal of Family Issues, 27,* 1561–1582.

McCall, R. B. (1977). Challenges to a science of developmental psychology. *Child Development, 48,* 333–344.

Overton, W. L. (2006). Developmental psychology: Philosophy, concepts, and methodology, In R. M. Lerner (Ed.), *Handbook of child psychology: Vol. 1. Theoretical models of human development* (5th ed., pp. 18–88). New York: Wiley.

Parkin, C. M., & Lataigne, M. (2007, June). *Adolescents' perspectives on parental rules.* Paper presented at the Annual Meeting of the Canadian Psychological Association, Calgary, Alberta, Canada.

Patterson, G. R. (1982). *Coercive family process.* Eugene, OR: Castalia.

Patterson, G. R. (1997). Performance models for parenting: A social interactional perspective. In J. E. Grusec & L. Kuczynski (Eds.), *Parenting and the internalization of values: A handbook of contemporary theory* (pp. 193–226). New York: Wiley.

Patterson, G., Reid, J., & Dishion, T. (1992). *Antisocial boys: Vol. 4. A social interactional approach.* Eugene, OR: Castalia.

Reigel, K. F. (1976). The dialectics of human development. *American Psychologist, 10,* 689–700.

Rubin, K. H., Nelson, L. J., Hastings, P., & Asendorpf, J. (1999). The transaction between parents' perceptions of their children's shyness and their parenting styles. *International Journal of Behavioral Development, 23,* 937–957.

Sameroff, A. (1975a). Early influences on development: Fact or fancy? *Merrill-Palmer Quarterly, 21,* 267–293.

Sameroff, A. (1975b). Transactional models of early social relations. *Human Development, 18,* 65–79.

Sameroff, A. J., & MacKenzie, M. J. (2003). Research strategies for capturing transactional models of development: The limits of the possible. *Development and Psychopathology, 15,* 613–640.

Sears, R. R. (1951). A theoretical framework for personality and social behavior. *American Psychologist, 6,* 476–482.

Shantz, C. U. (1987). Conflicts between children. *Child Development, 58,* 283–305.

Shantz, C. U., & Hartup, W. W. (1992). *Conflict in child and adolescent development.* Cambridge, England: Cambridge University Press.

Stern, M., & Hildebrandt, K. A. (1986) Prematurity stereotyping: Effects of mother–infant interaction. *Child Development, 57,* 308–315.

Trickett, P. K., & Kuczynski, L. (1987). Children's misbehavior and parental discipline strategies in abusive and nonabusive families. *Developmental Psychology, 22,* 21–44.

Trommsdorff, G., & Kornadt, H. (2003). Parent–child relations in cross-cultural perspective. In L. Kuczynski (Ed.), *Handbook of dynamics in parent–child relations* (pp. 271–306). Thousand Oaks, CA: Sage.

Valsiner, J. (1989). *Human development and culture*. Lexington, MA: Heath.

Valsiner, J. (2000). *Culture and human development*. London: Sage.

Valsiner, J. (2006). Ambivalence under scrutiny: Returning to the future. *Estudios de Psicologia, 27*, 117–130.

Valsiner, J., & Connolly, K. J. (2003). The nature of development: The continuing dialogue of processes and outcomes. In J. Valsiner & K. J. Connolly (Eds.), *Handbook of developmental psychology* (pp. xi–xviii). Thousand Oaks, CA: Sage.

Vulchinch, S. (1987). Starting and stopping spontaneous family conflicts. *Journal of Marriage and the Family, 49*, 591–601.

What the heck is dialectics? (n.d.) Retrieved December 10, 2008, from http://home.igc.org/~venceremos/whatheck.htm

Youniss, J., & Smollar, J. (1984). *Adolescent relations with mothers, fathers, and friends*. Chicago: University of Chicago Press.

V

AFTERWORD

14
WHAT IS A TRANSACTION?

ALAN FOGEL

In the concluding chapter of this book on transactional processes in development, my goal is to collect the previous chapters under an organizing theme and then to describe my own perspective on transactional processes. From my perspective as a reader of these chapters and as an informed participant-observer in the field of transactional research on interpersonal communication and development for the past 35 years, the theme that emerges from reading the previous chapters and the original writings of Sameroff and colleagues is, "What is a transaction?"

The original theoretical writings on the transactional approach by Sameroff and colleagues were sufficiently provocative to inspire a good deal of work and an impetus to thinking beyond the bidirectional model. Transaction implies, at a minimum, an ongoing process of mutual and emergent effects within relationships with a view to how these processes contribute to the formation of different developmental pathways. From the beginning, transactions were more than just mutual effects; they included the concept of transformation. The transactional approach, however, was framed to be sufficiently general to allow for a wide range of research tactics, as illustrated by the chapters of this book.

One tactic for studying transactions is to focus on the mutual effects between participants by using summary measures in longitudinal data. In this case, the variables are composite measures of each person in the relationship, such as from self-report or observational measures (e.g., externalizing, aggressive, warm, shy). This is the approach used in the chapters by Fontaine and Dodge (chap. 7); Bornstein (chap. 8); Bugental (chap. 6); Gershoff, Aber, and Clements (chap. 11); Shaw, Gross, and Moilanen (chap. 5); MacKenzie and McDonough (chap. 3); Morrison and Connor (chap. 10); and Gauvain (chap. 9). Different multivariate and time-based statistical strategies are used in these approaches, including latent growth curve modeling and structural equation modeling. These approaches allow for the examination of individual cases against the background of the larger group.

Findings from these studies all reveal in some way that the cointeracting participants influence each other over time. Morrison and Conner (chap. 10), for example, find that the quality of first-grade instruction alters the impact of instruction on the child in later grades because of individual differences in math and reading skills acquired in the first grade. It is not that first-grade instruction is a simple predictor of later school outcomes but that what happens in first grade appears to set off a chain of interlocking effects between the child and the teacher in later grades, an ongoing pattern of interaction that follows a particular developmental trajectory toward high or low achievement. A similar long-term pattern can be seen in the patterns of maternal depression and child conduct problems over time, each person's behavior continuing to affect the other in a mutual impasse that sustains each person's symptoms for long periods (Shaw et al., chap. 5).

Fontaine and Dodge (chap. 7) discuss the emergence of stable patterns of childhood aggression over time, sustained by the child's evaluation and selection of behavior that is most likely to get a desired response from peers. Bugental (chap. 6) talks about how parents who perceive themselves as having little power, coupled with a difficult child, create mutually amplifying patterns over time—in part sustained by the psychophysiology of threat and fear. Bornstein (chap. 8) proposes a model of parent–child specificity by which particular matchups, in the context of cultural beliefs, send dyads down different developmental pathways through transactional processes. Gauvain (chap. 9) takes this model one step further to speculate about how cultures themselves can change over time and follow particular developmental trajectories as the result of intergenerational transactions on the one hand and peer group transactions on the other.

In relation to the question "What is a transaction?", my sense is that none of the authors would adopt the view that the transaction is the particular set of correlations between the summary variables in their statistical models. These statistical models only hint at what the authors presume to be the underlying transactions by showing, at least in a macroscopic way, that partners in ongoing relationships affect each other and that these mutual effects lead to the consolidation of a particular developmental trajectory. Several of these authors (Bugental, chap. 6; Fontaine & Dodge, chap. 7; Shaw et al., chap. 5) present the results of clinical interventions that appear to shift these trajectories away from pathological pathways into more adaptive ones. This approach is another important, yet still indirect, way of implying the centrality of mutual transactions in shaping developmental outcomes.

So, if these statistical models of summary variables and the clinical outcome studies only suggest or hint at transactions, then it still leaves open the question, "What is a transaction?" The authors of chapters 1–2 and 3–10 all seem to think that transactions take place in real time, the moment-to-moment interplay between, for example, a hesitant mom and a child who can perceive the hesitancy and use aggressive behavior for instrumental gain, which leads the parent to back off further and amplifies the child's tendency to hurl insults or threats or smash the mother's possessions. The authors all also profess the difficulty of studying that real-time process, leaving that to speculation, informal observation, or to other investigators.

The study of these real-time communication processes does not necessarily answer the question, "What is a transaction?" If one is thinking about each person as emitting discrete actions in real time that lead to some convergence of an interaction pattern—for example, mutual threats leading to a full-blown argument—this is structurally the same as the discrete summary variables measured from each participant at different ages in a longitudinal study, which lead to differing outcome pathways (e.g., an aggressive child coupled with a compliant mother). Thinking about such discrete processes, whether in real or developmental time, gets one no closer to answering the question, "What is a transaction?"

A wide variety of statistical and simulation models exist to conceptualize and analyze real-time discrete sequences. These tools have continued to become more mathematically sophisticated, capable of describing complex lead–lag relationships, timing factors in interaction, and other quantitative indices of social engagement and sensitivity (cf. Jaffe, Beebe, Feldstein, Crown, & Jasnow, 2001).

The other authors in this book—Gershoff et al. (chap. 11), Gonzalez (chap. 12), Kuczynski and Parkin (chap. 13), and Olson and Lunkenheimer (chap. 4)—have taken on this more daunting task of studying the communication process in real time. Each of these authors recognizes that the discrete state model of communication—he says, she says—is not sufficient to constitute a transaction. These discrete state models, so-called *bidirectional* models, miss the heart of real-time communication by packaging a much more dynamic process into artificial elements.

Gershoff et al. (chap. 11) and Gonzales (chap. 12) take a more statistical approach, whereas Kuczynski and Parkin (chap. 13) and Olson and Lunkenheimer (chap. 4) take up the issue from a more qualitative perspective. All these authors assert that transactions involve a continuous and ongoing process. *Continuous* means that there is an opportunity for partners to influence each other simultaneously and not simply in a linear sequence of cause and effect. One important step in this direction is the inclusion of simultaneous mutual influences in statistical models such as the continuous time cross-lagged panel models used by Gershoff et al.

When one conceptually admits that there is a continuous and simultaneous mutual influence, one buys a ticket out of the territory of most of the existing scientific models and metaphors that imply control, well-behaved (stable, reliable, valid) measurement units, statistical models with straight lines, and arrows representing simple and mediated causes and effects. What is the other territory to which one travels? It is one in which it is more difficult to define and control variables, some would say impossible and fundamentally indeterminate (Fogel, 2006). It is a land in which qualitative transformations emerge that cannot be predicted by a simple concatenation of prior sequences of events. It is a place in which the old methods do not apply and in which new navigational tools need to be invented.

Olson and Lunkenheimer (chap. 4) cite theory and research that has assumed a continuous and ongoing process of communication including concepts such as coregulation (Fogel, 1993), mutual regulation (Tronick, 1998), and dyadic synchrony (cf. Hubbard, Dodge, Cillessen, Coie, & Schwartz, 2001; Skuban, Shaw, Gardner, Supplee, & Nichols; 2006). In some cases, these scholars have used dyadic assessments of synchrony and coregulation rather than assessing each individual's contribution. When people affect each other in an ongoing manner, one quickly loses the ability to track what an individual is doing compared with seeing the dyadic pattern.

The continuous process model of communication (Fogel, 1993) suggests that communication is more than the sum of each individual contribution but rather is created and emergent from the communication process. According to Hinde (1985), with regard to animal postures of aggression, "Such signals are thus to be seen as involving negotiation with the rival as well as an expression of internal state. The term *negotiation* does not necessarily imply manipulation but emphasizes the continuous interaction between the two individuals involved" (Hinde, 1985, p. 111).

No behavior emerges fully formed from the brain without the ongoing influences at the periphery of the body and in the local environment in the very act of formation (Thelen & Smith, 1994). A continuous process model follows from a dynamic systems perspective (Fogel, 1993; Kelso, 2000; Thelen & Smith, 1994) in which there are self-organizing patterns of coactivity in the communication system called *attractors*. These attractors, though patterned, are nevertheless fluid and dynamic and leave room for emergence of novelty through coregulation processes. Through self-organization, all the components of a system are mutually affecting each other in such a way that they temporarily lose their individual identities, thereby forming cooperative units, or coordinative structures, that have unique properties transcending the individual components. Gonzales (chap. 12) proposes a number of novel statistical methods that move away from the traditional dependent–independent variable dichotomies and work from models of simultaneous and ongoing influences in which the variables, and their influence on each other, change continuously (see also Lewis, Lamey, & Douglas, 1999; Ryan, Gottman, Murray, Carrère, & Swanson, 2000; van Geert, 1998).

Kuczynski and Parkin (chap. 13) provide perhaps the most detailed map of this new territory. They argue against linear thinking, proposing to view individuals as not fully formed but as continuously changing in a dialectical process of mutual engagement. They propose that the relationship, rather than the individuals, is the appropriate unit of analysis. Most important, however, is their recognition that transactions are not only simultaneous and ongoing but also transformational and emergent. "A transaction involves a turning point or qualitative shift" (p. 260). They recommend that researchers focus not on the linear causes of these transformations but rather on their dynamics of timing and quality. Are the transformations abrupt (like a sudden parental divorce) or gradual (as in an effective therapeutic relationship)?

Paradoxically, this leads away from what appear to be the mutual exchanges of behavior—what many have assumed to be the essence of transactions—toward a focus on system-level qualities (engaged vs. unengaged, mutually amplifying vs. mutually regulating, sudden vs. gradual change) as a way to study transactional processes. This makes little sense unless one acknowledges that the answer to the question "What is a transaction?" is located in that strange and new territory and not in the old and well-mapped territory. The answer is not in mutual effects, because these descriptions presuppose an individual who emits behaviors that affect and are effects of other's behaviors.

What matters is how to understand the process by which mother and son reach an impasse in communication only resolved by hostility, and more important, how to coax this entrenched pattern into a more loving and respectful one. It matters very little how many insults are exchanged because the impasse can sometimes flare up silently and only by brief glares and retreats. The process has created a dynamic self-sustaining system in which each individual is cocreated by the other and each is a slave to the relationship.

What is a transaction? I cannot define it because of the fluidity, transformation, and emergence of the real and developmental time dynamics. I cannot put people or their behavior into nicely packaged boxes. My career has been devoted to getting some footing in the shifting sands of this new territory without, I hope, giving in to easier and simpler payoffs (Fogel, 1993).

However, I can make concrete suggestions for how to study transactions and perhaps even how to change stable patterns (cf. Fogel, 1993, 2006; Fogel, Garvey, Hsu, &West-Stroming, 2006). And I can provide some different types of research and clinical questions to ask regarding the laws of change and the dynamics of transformation in relationship systems. How much can a system "take" before it breaks down? What is the process by which transformation occurs? Why do some individuals and systems seem to resist change and others embrace change? What are the factors that expand or limit the possibility for qualitative change in the system from an old pattern into a new one?

FRAMES

My colleagues and I (Fogel et al., 2006) have been studying recurring patterns of communication called *frames*. Frames are dynamically stable transactional patterns in ongoing relationships such as rituals, routines, and ways of doing things together. Frames recur repeatedly over weeks and months and are reconstituted dynamically and dyadically on each reappearance. Frames reflect a higher level pattern that encompasses a complex interlocking network of moment-to-moment shifts in coaction between participants, a transactional pattern.

From a developmental perspective, frames typically last for only a finite period in an interindividual relationship. In a romantic partnership, for example, engagement or betrothal is a frame that has a limited time of life, between courtship and marriage. Particular play or caregiving or disciplinary routines have a period of ascendancy, a peak, and a decline over time in a parent–child relationship.

We studied this process of developmental change in frames by weekly videotaping of a mother and infant playing together over a 4-month observation period when the infant was between 2 and 6 months of age, during which there is a known developmental transition from the mother's guidance of infant object play to the mother's support of infant independent object play. To capture the details of this transition, we used weekly observations in a microgenetic research design in which the interobservation period is relatively brief compared with the developmental change interval that occurs over several months (Fogel et al., 2006).

From the videos, we identified four mutually exclusive and exhaustive relationship frames. The *social frame* was coded when the topic of communication was face-to-face play without objects. The *guided object frame* was coded when the mother took an active role in demonstrating and scaffolding the infant's use of objects. The *nonguided object frame* was coded when the infant played with objects without the mother's direct assistance but with her ongoing attention and verbal commentary. The *social object mixed frame* was coded when elements both of face-to-face play and guided object play appeared at the same time, as when the mother used a toy to touch the infant's face or body while vocalizing in an expressive manner typical of the social frame.

We then computed the durations of each of the frames and found that the durations of the social and the guided object frames declined over this period, the nonguided object frame increased in duration, and the social object mixed frame had an inverted-U shape. This same developmental pattern was found in all but two of our dyads. There was an early and a late stable period in which the durations of frames remained reasonably constant. In between, we typically found a transition period with a "microgenetic signature" of quantitative change characteristics for a developmental transition:

- There was an increase in week-to-week variability of the frames.
- One frame, in this case the social object mixed frame, became predominant as the previously dominant frame (guided object frame) was declining and as the newly emerging frame (nonguided object frame) was increasing.

The frame that had the inverted-U shape we called the *bridging frame* because it appeared to serve as an intermediary, a pathway for change, that helped the system move from the old routine into the new one. The engagement or betrothal period between courtship and marriage is an example of a bridging frame in a romantic relationship. Divorcing couples may have a period of separation that serves as a bridge to a final settlement.

Using the example of a divorcing couple, as my colleagues and I have done (Fogel et al., 2006), one can respond at many levels from the perspective of "What is a transaction?" There are the transactions of moment-to-moment coordinations within each frame. These transactions actually change over time. What goes on during the guided object frame at Week 2 is different from what goes on at Week 7, as we documented using qualitative analysis (Fogel et al., 2006). Part of what accounts for this change are the transactions between the different frames in the relationship at any given point in time. Couples make real-time transitions between frames, and part of what regulates this process is the state of each frame's development and its relative availability to the couple.

The frames transact with each other by a number of processes as part of a relationship system that is itself developing over time. For example, there is a process of *permeability* in which frames blend into each other during the transition period. Another process is *recapitulation*, which is similar to the old concept of regression but at the level of the relationship. The couple seems to bring back, but only temporarily, an old frame or pattern of frames that was useful and familiar in the past. A final process that we observed during the developmental transition period is the emergence of spontaneous *innovations*—sudden leaps of novelty, new ways of relating, and sparks of inspiration in the dyad as if, caught up in the felt inevitability of a dynamic change process, the couple could experiment with new ways of doing things.

Developmental transitions, then, include a variety of processes with transactional characteristics described previously as bridging, permeability, recapitulation, and innovation. These processes arise in some way to move the dyad from the beginning of the transition to the end, and there is a complex, nested set of transactions between events in real time, between frames in real time, and between frames and events over developmental time.

These findings have clear clinical implications. One way to think about trauma, for example, is in terms of the absence of bridging. When events occur too rapidly for this process to unfold (accidents, war, assault, abuse) and for individuals' psychophysiology to adapt—in other words, when there is not an opportunity for bridging—one is likely to see symptoms of posttraumatic stress disorder and other related disorders. For sufferers of trauma, therapeutic bridging (along with opportunities for recapitulation, innovation, and permeability) may be crucial to recovery.

Table 14.1 shows the findings from a study of treatment for eating disorders (Cockell, Zaitsoff, & Geller, 2004). I have recast those findings in terms of the bridging model. Those women who maintained the contacts with the past supports (Period 1) and developed emerging frames for the future (Period 3), while bridging these two frames with concrete application of program skills (Period 2), were most likely to recover. This study was actually retrospective rather than microgenetic. If microgenetic methods were applied to treatment research, one might find all the developmental processes mentioned here (e.g., temporary recapitulation to older ways of eating as one attempts to give them up but before a new and stable way of relating to food has emerged). Put

TABLE 14.1
Characteristics of Women Who Were Successful or Unsuccessful in Recovering From Eating Disorders as a Result of a Treatment Program

Period	Successful recovery	Unsuccessful recovery
1. Historical frame	Maintaining connection to existing social supports	Sense of disconnection, lack of social support
2. Bridging frame	Using program skills in everyday life	Self-defeating beliefs, failure to apply program skills
3. Emerging frame	Focusing beyond the eating disorder (i.e., making changes in their lives and developing "higher" values)	Inability to maintain diet, focus on immediate soothing of stress rather than setting longer term goals for the future

Note. Some data are from Cockell et al. (2004).

another way, and coming back to the issues raised at the end of the previous section, psychologists can diagnose a change process not by its microscopic exchanges of actions and not by correlations across individuals but rather by the systemwide features of within-case microgenetic change described here.

I end by repeating the series of questions that need answers to amplify and perhaps reframe the transactional model: How much can a system take before it breaks down? What is the process by which transformation occurs? Why do some individuals and systems seem to resist change and others embrace change? What are the factors that expand or limit the possibility for qualitative change in the system from an old pattern into a new one? These are all questions that may be addressed and answered when the transactional model is applied to microgenetic change processes.

REFERENCES

Cockell, S., Zaitsoff, S., & Geller, J. (2004). Maintaining change following eating disorder treatment. *Professional Psychology: Research and Practice*, *35*, 527–534.

Fogel, A. (1993). *Developing through relationships*. Chicago: The University of Chicago Press.

Fogel, A. (2006). Dynamic systems research on interindividual communication: The transformation of meaning making. *Journal of Developmental Processes*, *1*, 7–30.

Fogel, A., Garvey, A., Hsu, H., & West-Stroming, D. (2006). *Change processes in relationships: A relational–historical research approach*. Cambridge, England: Cambridge University Press.

Hinde, R. A. (1985). Expression and negotiation. In G. Zivin (Ed.), *The development of expressive behavior: Biology–environment interactions* (pp. 103–116). New York: Academic Press.

Hubbard, J., Dodge, K., Cillessen, A., Coie, J., & Schwartz, D. (2001). The dyadic nature of social information processing in boys' reactive and proactive aggression. *Journal of Personality and Social Psychology*, *80*(2), 268–280.

Jaffe, J., Beebe, B., Feldstein, S., Crown, C. L., & Jasnow, M. (2001). Rhythms of dialogue in infancy. *Monographs of the Society for Research in Child Development*, *66*(2, Serial No. 265).

Kelso, J. A. S. (2000). Principles of dynamic pattern formation and change for a science of human behavior. In L. R. Bergman, R. B. Cairns, L. Nilsson, & L. Nystedt (Eds.), *Developmental science and the holistic approach* (pp. 63–83). Mahwah, NJ: Erlbaum.

Lewis, M., Lamey, A. V., & Douglas, L. (1999). A new dynamic systems method for the analysis of early socioemotional development. *Developmental Science*, *2*, 457–475.

Ryan, K. D., Gottman, J. M., Murray, J. D., Carrère, S., & Swanson, C. (2000). Theoretical and mathematical modeling of marriage. In M. D. Lewis &

I. Granic (Eds.), *Emotion, development, and self-organization* (pp. 349–372). New York: Cambridge University Press.

Skuban, E., Shaw, D., Gardner, F., Supplee, L., & Nichols, S. (2006). The correlates of dyadic synchrony in high-risk, low-income toddler boys. *Infant Behavior & Development, 29*(3), 423–434.

Thelen, E., & Smith, L. B. (1994). *A dynamic systems approach to the development of cognition and action*. Cambridge, MA: MIT Press.

Tronick, E. Z. (1998). Dyadically expanded states of consciousness and the process of therapeutic change. *Infant Mental Health Journal, 19*, 290–299.

van Geert, P. (1998). We almost had a great future behind us: The contribution of non-linear dynamics to developmental-science-in-the-making. *Developmental Science, 1*, 143–159.

INDEX

ABOUT THE EDITOR

Arnold Sameroff, PhD, a developmental psychologist, is currently professor of psychology at the University of Michigan, where he is also chair of the Developmental Psychology Graduate Training Program and director of the Development and Mental Health Research Program at the Center for Human Growth and Development. His influential theoretical work on ecological transactional models of development has helped to move researchers to more dynamic, system-based research efforts for understanding healthy child development, and his research on environmental risk and promotive factors has fostered a more comprehensive understanding of what is necessary to improve the cognitive and social–emotional welfare of children. Among the high-risk groups he is currently studying are infants with physiologic regulatory problems, children with depressed parents, adolescents living in low-resource neighborhoods, and adults reared in families with parental mental illness. He has published numerous research articles and 12 books and monographs, including *The Five to Seven Year Shift: The Age of Reason and Responsibility*; *Handbook of Developmental Psychopathology*; *Treating Early Relationship Problems: Infant, Parent, and Interaction Therapies*; and, with Sheryl Olson, the forthcoming *Regulatory Processes in the Development of Behavior Problems: Biological, Behavioral, and Social–Ecological Interactions*. Among his honors are the

Distinguished Scientific Contributions Award from the Society for Research in Child Development and the G. Stanley Hall Award from Division 7 (Developmental Psychology) of the American Psychological Association. He is former president of Division 7 of the American Psychological Association and the International Society for Infant Studies and is current president of the Society for Research in Child Development.